ANALOG-BASEBAND ARCHITECTURES AND CIRCUITS FOR MULTISTANDARD AND LOW-VOLTAGE WIRELESS TRANSCEIVERS

ANALOG CIRCUITS AND SIGNAL PROCESSING SERIES

Consulting Editor: Mohammed Ismail. *Ohio State University*

Titles in Series:

ANALOG-BASEBAND ARCHITECTURES AND CIRCUITS FOR MULTISTANDARD AND LOW-VOLTAGE WIRELESS TRANSCEIVERS
Pui-In Mak, Seng-Pan U, Rui Paulo Martins
ISBN: 978-1-4020-6432-6

ULTRA LOW POWER CAPACITIVE SENSOR INTERFACES
Bracke, W., Puers, R. (et al.)
ISBN: 978-1-4020-6231-5

BROADBAND OPTO-ELECTRICAL RECEIVERS IN STANDARD CMOS
Hermans, C., Steyaert, M.
ISBN: 978-1-4020-6221-6

CMOS MULTI-CHANNEL SINGLE-CHIP RECEIVERS FOR MULTI-GIGABIT OPT…
Muller, P., Leblebici, Y.
ISBN: 978-1-4020-5911-7

SWITCHED-CAPACITOR TECHNIQUES FOR HIGH-ACCURACY FILTER AND ADC…
Quinn, P.J., Roermund, A.H.M.v.
ISBN: 978-1-4020-6257-5

LOW-FREQUENCY NOISE IN ADVANCED MOS DEVICES
von Haartman, M., Östling, M.
ISBN: 978-1-4020-5909-4

CMOS SINGLE CHIP FAST FREQUENCY HOPPING SYNTHESIZERS FOR WIRELESS MULTI-GIGAHERTZ APPLICATIONS
Bourdi, Taoufik, Kale, Izzet
ISBN: 978-1-4020-5927-8

ANALOG CIRCUIT DESIGN TECHNIQUES AT 0.5V
Chatterjee, S., Kinget, P., Tsividis, Y., Pun, K.P.
ISBN-10: 0-387-69953-8

IQ CALIBRATION TECHNIQUES FOR CMOS RADIO TRANCEIVERS
Chen, Sao-Jie, Hsieh, Yong-Hsiang
ISBN-10: 1-4020-5082-8

FULL-CHIP NANOMETER ROUTING TECHNIQUES
Ho, Tsung-Yi, Chang, Yao-Wen, Chen, Sao-Jie
ISBN: 978-1-4020-6194-3

THE GM/ID DESIGN METHODOLOGY FOR CMOS ANALOG LOW POWER INTEGRATED CIRCUITS
Jespers, Paul G.A.
ISBN-10: 0-387-47100-6

PRECISION TEMPERATURE SENSORS IN CMOS TECHNOLOGY
Pertijs, Michiel A.P., Huijsing, Johan H.
ISBN-10: 1-4020-5257-X

CMOS CURRENT-MODE CIRCUITS FOR DATA COMMUNICATIONS
Yuan, Fei
ISBN: 0-387-29758-8

RF POWER AMPLIFIERS FOR MOBILE COMMUNICATIONS
Reynaert, Patrick, Steyaert, Michiel
ISBN: 1-4020-5116-6

ADVANCED DESIGN TECHNIQUES FOR RF POWER AMPLIFIERS
Rudiakova, A.N., Krizhanovski, V.
ISBN: 1-4020-4638-3

CMOS CASCADE SIGMA-DELTA MODULATORS FOR SENSORS AND TELECOM
del Río, R., Medeiro, F., Pérez-Verdú, B., de la Rosa, J.M., Rodríguez-Vázquez, A.
ISBN: 1-4020-4775-4

SIGMA DELTA A/D CONVERSION FOR SIGNAL CONDITIONING
Philips, K., van Roermund, A.H.M.
Vol. 874, ISBN: I-4020-4679-0

CALIBRATION TECHNIQUES IN NYQUIST A/D CONVERTERS
van der Ploeg, H., Nauta, B.
Vol. 873, ISBN: 1-4020-4634-0

ADAPTIVE TECHNIQUES FOR MIXED SIGNAL SYSTEM ON CHIP
Fayed, A., Ismail, M.
Vol. 872, ISBN: 0-387-32154-3

WIDE-BANDWIDTH HIGH-DYNAMIC RANGE D/A CONVERTERS
Doris, Konstantinos, van Roermund, Arthur, Leenaerts, Domine
Vol. 871, ISBN: 0-387-30415-0

METHODOLOGY FOR THE DIGITAL CALIBRATION OF ANALOG CIRCUITS AND SYSTEMS: WITH CASE STUDIES
Pastre, Marc, Kayal, Maher
Vol. 870, ISBN: 1-4020-4252-3

HIGH-SPEED PHOTODIODES IN STANDARD CMOS TECHNOLOGY
Radovanovic, Sasa, Annema, Anne-Johan, Nauta, Bram
Vol. 869, ISBN: 0-387-28591-1

ANALOG-BASEBAND ARCHITECTURES AND CIRCUITS FOR MULTISTANDARD AND LOW-VOLTAGE WIRELESS TRANSCEIVERS

Pui-In Mak

University of Macau, China

Seng-Pan U

University of Macau and Chipidea Microelectronics (Macau), Ltd., China

Rui Paulo Martins

University of Macau, China, and Technical University of Lisbon, Portugal

Springer

A C.I.P. Catalogue record for this book is available from the Library of Congress.

ISBN 978-90-481-7640-3
ISBN 978-1-4020-6433-3 (e-book)

Published by Springer,
P.O. Box 17, 3300 AA Dordrecht, The Netherlands.

www.springer.com

Printed on acid-free paper

This book is dedicated to

Our Families

Contents

Preface

The prospect of initializing a network-ubiquitous society in the years to come has led to the development of multistandard-compliant wireless transceivers for seamless roaming among multiple networks. To ensure a commercial success of such a development, the manufacturing cost and power consumption of the system chips have to be minimized. The use of an advanced technology and a high level of integration have continued to be the most effective ways for cost and power minimization, given that wireless chips integrate large amounts of digital logic for computation. Regrettably, entering into the nanoelectronics era, the thinner transistor gate oxide implicates great challenges in the design of the analog front-ends. While a low-voltage supply is imposed to maintain device reliability, a relatively large threshold voltage is also necessitated to limit the leakage current. Thus, transceiver architectures and circuits which will befit future full integration of multistandard wireless transceivers in sub-1V nanoscale CMOS processes must be highly reconfigurable and robustly operational underneath a low-voltage supply.

This book presents novel analog-baseband architectures and circuits that help realizing multistandard and low-voltage wireless transceivers. The main contents are presented from Chapter 2 to Chapter 6, as pictorially outlined in Figure 1.

- **Chapter 1** overviews the current wireless-IC developments and presents the motivation and research objectives of this book.

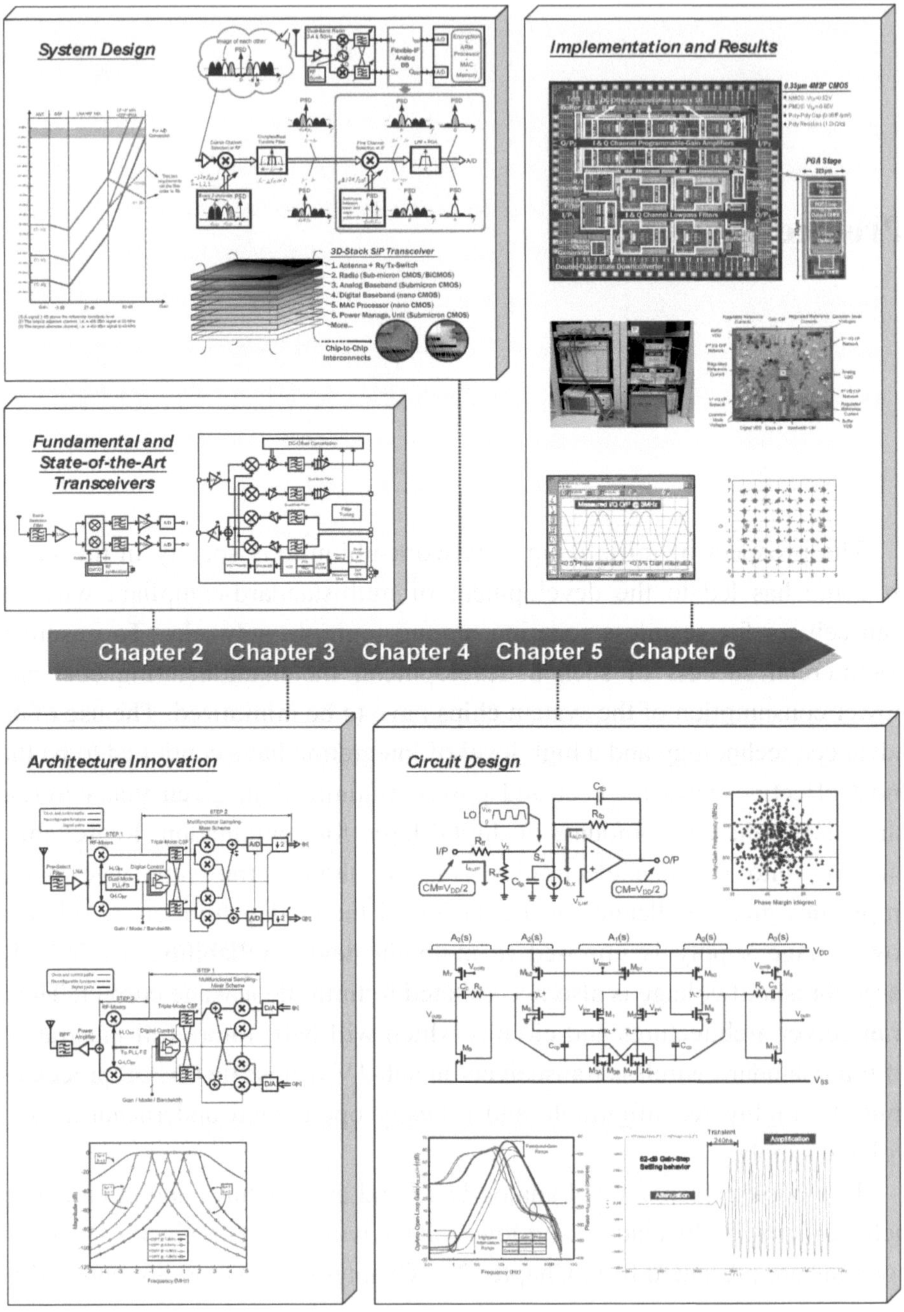

Figure 1. Outline of the book

- **Chapter 2** studies the fundamental receiver and transmitter architectures and reviews the physical layer (PHY) specifications of modern wireless communication standards. A statistical summary (with 100+ references) of most exploitation transmitter, receiver and transceiver architectures for modern wireless communication systems has been surveyed. The references are collected from the solid-state circuit forums: ISSCC, CICC, VLSI and ESSCIRC, from 1997 to 2005. The final part presents case studies of the state-of-the-art multistandard receivers and transceivers. Their achieved performances are compared in relation to their architectural choices and levels of block sharing.

- **Chapter 3** introduces a novel coarse-RF fine-IF (two-step) channel-selection technique. Through the reconfiguration of receiver and transmitter analog basebands, it enables not only relaxation of the RF frequency synthesizer's and local oscillator's design specifications, but also an efficient multistandard compliance by synthesizing the low-IF/ zero-IF mode in the receiver; and the direct-up/two-step-up mode in the transmitter. The reconfiguration is mainly contributed by a triple-mode channel-select filter and a multifunctional sampling-mixer scheme.

- **Chapter 4** presents the system design of a system-in-a-package (SiP) receiver analog baseband for IEEE 802.11a/b/g WLAN. It not only has the proposed two-step channel-selection technique reinforced, but also features globally a 3D-stack SiP floorplan for testability and routability, a flexible-IF (low-IF/zero-IF) reception capability and an area-efficient baseband-signal conditioning approach for different modes of operation.

- **Chapter 5** deals with the functional-block design of the above-mentioned receiver analog baseband with emphasis on reconfigurability, low-voltage supply, power and area savings. Methodical description of low-voltage circuit subblocks is given first, followed by the presentation of three novel functional blocks:

  1. A double-quadrature-downconversion filter based on a *series-switching* mixer-quad realizes a wideband mismatch-insensitive I/Q demodulation while offering the capabilities of clock-rate-defined IF reception, channel-selection filtering and sideband selection.
  2. A *switched-current-resistor* programmable-gain amplifier provides a transient-free constant-bandwidth gain adjustment.

3. An *inside-OpAmp* dc-offset canceler saves the silicon area required for realizing a large time constant on chip while maximizing its highpass-pole switchability for fast settling in dc-offset transients.

- **Chapter 6** summarizes the design issues, test strategy and experimental results of a low-voltage receiver analog-baseband IC for IEEE 802.11a/b/g WLAN. It is based on the system described in Chapter 4 and the low-voltage functional blocks presented in Chapter 5. A special test strategy enabled precision building-block and full subsystem characterization is also proposed.

- **Chapter 7** draws the relevant concluding remarks and benchmarks the position of the described works to the state of the art. Timely future developments extending the current research are also suggested.

Pui-In Mak
Seng-Pan U
Rui Paulo Martins

Aknowledgments

We would like to thank Professor Kenneth W. Martin (University of Toronto), Professor Peter Chung-Yu Wu (National Chiao-Tung University), Professor Zhihua Wang (Tsinghua University), Professor Vai-Pan Iu (University of Macau) and Professor Ying-Duo Han (University of Macau) for examining the Ph.D. thesis of Pui-In Mak that is the basis of this book.

We acknowledge the Research Committee of University of Macau for the financial support, and the Macau Science and Technology Development Fund for funding a US patent application of this research.

We are very grateful to our colleagues, especially Ka-Hou Ao Ieong, in the Analog and Mixed-Signal VLSI Laboratory for fruitful discussions.

Finally, we send our heartfelt appreciation to our families, who endured our dedication to this book.

Acknowledgements

List of Abbreviations

ΣΔ	:	Sigma-Delta
1/*f*	:	Flicker Noise
2SU	:	Two Step Up
3D	:	3-Dimensional
3G/4G	:	3rd/4th Generation
A-BS	:	Analog-Baseband Sampling
AC	:	Alternating Current
ACPR	:	Adjacent Channel Power Ratio
ADC (A/D)	:	Analog-to-Digital Converter
A-DQS	:	Analog Double-Quadrature Sampling
AFE	:	Analog Front-End
AMPS	:	Advanced Mobile Phone System
ANT	:	Antenna
AWGN	:	Additive white Gaussian noise
BB	:	Baseband
BiCMOS	:	Bipolar Complementary Metal Oxide Semiconductor
BPF	:	Bandpass Filter
BSF	:	Band-Selection Filter
BUF	:	Buffer
BW	:	Bandwidth
CCK	:	Complementary Code Keying
CDMA	:	Code Division Multiple Access
CF	:	Complex Filter

CICC	:	Custom Integrated Circuits Conference
CG	:	Conversion Gain
CLKGEN	:	Clock Generator
CM	:	Common Mode
CMFB	:	Common Mode Feedback
CMOS	:	Complementary Metal Oxide Semiconductor
CMRR	:	Common-Mode Rejection Ratio
CQFP	:	Ceramic Quad Flat-Pack
CS	:	Channel Spacing
CSF	:	Channel-Selection Filter
CSSF	:	Complex-Signal Spectral-Flow
CT	:	Continuous-Time
CV	:	Capacitance-Voltage
DAC (D/A)	:	Digital-to-Analog Converter
DC	:	Direct Current
DCS	:	Digital Cellular System
D-DQS	:	Digital Double-Quadrature Sampling
DECT	:	Digital Enhanced Cordless Telecommunications
DOC	:	DC-Offset Canceler
DQDC	:	Double-Quadrature Downconverter
DQDF	:	Double-Quadrature-Downconversion Filter
DRC	:	Design Rule Check
DSP	:	Digital Signal Processor
DSSS	:	Direct-Sequence Spread Spectrum
DU	:	Direct Up
EDA	:	Electronics Design Automation
EDGE	:	Enhanced Data rate for GSM Evolution
EMI	:	Electromagnetic Interference
ENOB	:	Effective Number of Bits
ESD	:	Electrostatic Discharge
ESSCIRC	:	European Solid-State Circuits Conference
EVM	:	Error Vector Magnitude
FET	:	Field Effect Transistor
FFT	:	Fast Fourier Transform
FHSS	:	Frequency-Hopping Spread Spectrum
FS	:	Frequency Synthesizer

GBW	:	Gain-Bandwidth Product
GCA	:	Gain-Controllable Amplifier
GPS	:	Global-Positioning System
GSM	:	Global System for Mobile Communications
HPF	:	Highpass Filter
HD2/HD3	:	Second-/Third-order Harmonic Distortion
HIF	:	High Intermediate Frequency
I	:	In Phase
IC	:	Integrated Circuit
I-CMFB	:	Input Common-Mode Feedback
IEEE	:	Institute of Electrical and Electronics Engineering
IF	:	Intermediate Frequency
I/P	:	Input
IM3	:	Third-order Intermodulation Distortion
IP3	:	Third-order Intercept Point
IRR	:	Image-Rejection Ratio
ISM	:	Industrial, Scientific and Medical
ISI	:	Intersymbol Interference
ISSCC	:	International Solid-State Circuits Conference
ITRS	:	International Technology Roadmap for Semiconductors
I-to-V	:	Current-to-Voltage
LC	:	Inductor-Capacitor
LDO	:	Low-Dropout
LIF	:	Low IF
LNA	:	Low-Noise Amplifier
LO	:	Local Oscillator
LPF	:	Lowpass Filter
LV	:	Low Voltage
LVS	:	Layout versus Schematic
MAC	:	Medium Access Control
MEMS	:	Micro-Electro-Mechanical Systems
MIMO	:	Multiple-Input Multiple-Output
MIX	:	Mixer
MOSFET	:	Metal-Oxide Semiconductor Field-Effect Transistor
NF	:	Noise Figure
O-CMFB	:	Output Common-Mode Feedback

OFDM	:	Orthogonal Frequency-Division Multiplexing
O/P	:	Output
OpAmp	:	Operational Amplifier
OSR	:	Oversampling Ratio
OTA	:	Operational Transconductance Amplifier
PCB	:	Printed-Circuit Board
PCS	:	Personal Communication System
PER	:	Packet-Error Rate
PGA	:	Programmable-Gain Amplifier
PGC	:	Programmable-Gain Control
PHY	:	Physical Layer
PLL	:	Phase-Locked Loop
PM	:	Phase Margin
PN	:	Pseudorandom Noise
PSD	:	Power Spectrum Density
PSS	:	Power Supply Sensitivity
PVT	:	Process, Voltage and Temperature
Q	:	Quadrature Phase/Quality Factor
QAM	:	Quadrature Amplitude Modulation
QPSK	:	Quadrature Phase-Shift Keying
QVCO	:	Quadrature Voltage Control Oscillator
RC	:	Resistor-Capacitor
REF	:	Reference
R-to-I	:	Resistor-to-Current
RF	:	Radio Frequency
RFID	:	Radio Frequency Identification
RSSI	:	Receiver Signal Strength Indicator
RX	:	Receiver
SAW	:	Surface Acoustic Wave
SC	:	Switched Capacitor
SCR	:	Switched Current Resistor
SFDR	:	Spurious-Free Dynamic Range
S/H	:	Sample-and-Hold
SiGe	:	Silicon Germanium
SiP	:	System-in-Package
SMD	:	Surface-Mount Device

SNR	:	Signal-to-Noise Ratio
SoC	:	System-on-Chip
SS	:	Series Switching
SU	:	Superheterodyne
TDMA	:	Time Division Multiple Access
THD	:	Total Harmonic Distortion
TX	:	Transmitter
TXR	:	Transceiver
UWB	:	Ultra Wideband
VCM	:	Common Mode Voltage
VCO	:	Voltage Controlled Oscillator
VGA	:	Variable-Gain Amplifier
VLSI	:	Symposium on VLSI Circuits
V-to-I	:	Voltage-to-Current
WCDMA	:	Wideband Code Division Multiple Access
WLAN	:	Wireless Local Area Network
WPAN	:	Wireless Body Area Network
ZIF	:	Zero IF

Chapter 1

INTRODUCTION

1. EVOLUTION OF WIRELESS COMMUNICATIONS

Rich applications in handheld devices such as navigation and Internet access have witnessed the rapid evolution of wireless services from traditionally simple voice/text, to recently multimedia. As shown in Figure 1-1, the current wireless services cover a wide range of data rate related to distance, continuously trending toward offering the best connectivity. It will not be surprising that, in the 4th generation (4G), the portable devices will become a universal terminal [1.1]. From the viewpoint of integrated-circuit (IC) design, however, a one-fits-all "analog" front-end is undoubtedly thorny to be realized. The four dimensions of IC innovation, i.e., technology, devices, circuits and system architecture, will face unprecedented design challenges. This chapter will present a glimpse into them.

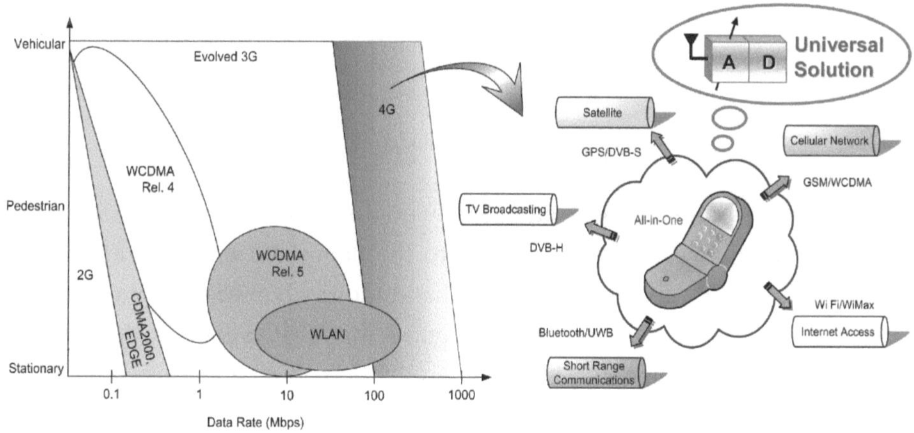

Figure 1-1. Evolution of wireless communications

2. WIRELESS-IC DESIGN CHALLENGES AND FUTURE PROSPECTS

One demanding feature of 4G wireless systems is that the terminal must be able to comply with more than one standard for best connectivity. GSM, GPS, WCDMA, WiFi (IEEE 802.11a/b/g/n), WiMax, Bluetooth, ZeeBee and Ultra Wideband (UWB), are certain prospective standards that can be considered as pieces in the puzzle of the final wireless-ubiquitous network. As a result, the manufacturing cost is determined critically by the efficiency in realizing the transceiver front-ends with high multistandardability, given that the covered standards may be distributed over a range of spectrum (Figure 1-2) and set very different specifications. Although the current 3rd-generation (3G) wireless systems have an improved data rate, high-speed signal transfers such as instantaneous video stream are not yet able to reach a quality of service (speed, liability and security, etc.) that is comparable with wireline.

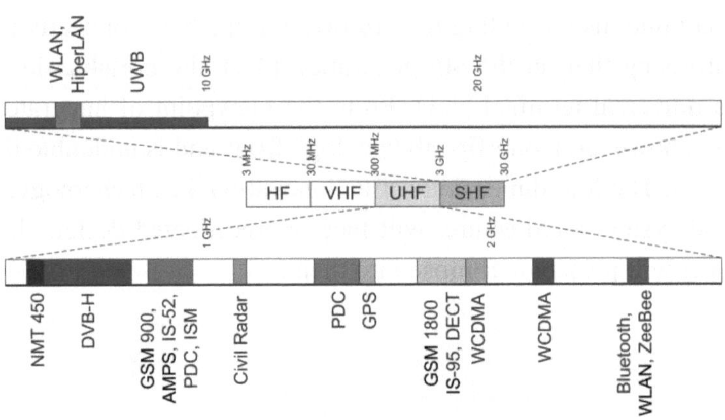

Figure 1-2. Spectrum usage of modern wireless standards

Providentially, advances in lithography continuously help maintaining the benefits of CMOS scaling. Entering into the nanoscale era, the fabrication materials of CMOS are being improved and more advanced process options and metal layers are being available, foreseeing CMOS still as the lowest-cost substrate of choice for massive production. Yet, there is simply no perfection; technology migration to nanoscale feature sizes brings along various new distressful limitations. On the one hand, the leakage (standby) current roughly doubles at every new technology node, imposing that the downscaling of threshold voltage will be decelerated, or even will have a halt, in the upcoming processes. The thinner gate oxide, on the other hand, imposes that the maximum supply voltage must be reduced consistently. A continuously shrinking supply-to-threshold-voltage ratio is hence anticipated, posing considerable challenges to the implementation. Moreover, lacks of accurate fine-linewidth transistor models, variability and signal integrity are other blockages that will derive the silicon from simulation results, making technology migration of large mixed-signal systems even harder to handle. Whether those issues are addressed or not will determine the successfulness of future wireless system-on-chip (SoC) or system-in-package (SiP) in nano-scale processes for continued economies of scaling.

To this end, diverse new possibilities have been undertaken to benefit from advanced technologies. Digital-assisted analog circuits, such as analog-to-digital (A/D) converters [1.2], have proved that they can achieve higher performances with low add-on cost because of the cheapness of linewidth digital circuits. Alternatively, most digital processors for wireless systems require certain analog front-ends to interface the real world. Such two examples imply that, up to some extent, purely analog and digital blocks will disappear and the trend would be towards complementarity rather than competition between both areas.

Heterogeneous technology domains other than simply analog and digital such as micro-electro-mechanical systems (MEMS) [1.3], microfluidic devices, optoelectronic and photonic devices, and other above-IC technologies [1.4][1.5][1.6] appear as decisive enablers that will possibly change the formation of future ICs and allow for very high-speed communication at extreme high frequency (e.g., D band, 110–170 GHz). New device architectures other than bulk MOSFET like SOI MOSFET, FinFET or tunneling FET are believed to be other key bottom-level enablers for the post-scaling of CMOS. In size miniaturization, squeezing heterogeneous chips with above-IC elements will eventually approach 3-D integration as the technology scaling hits its physical limits (perhaps at 5-nm node, the lithography scale will reach a few times atomic dimensions). This evolutionary repositioning will require, first, advanced electronic design automation (EDA) tools and design methodology allowing complete mixed-domain (e.g., electrical and mechanical) co-simulations, and second, substantial efforts on high-level macromodeling for simulation efficiency.

Most excitingly, it is believed that not counting cellular and WPAN/ WLAN networks, sensory systems based on impulse-based UWB techniques and Radio Frequency Identification (RFID) tags are other key technologies permitting a wireless-ubiquitous society connecting people. Biomedical, logistic, management, shopping, entertainment and many previously unthinkable products or services will be progressively converging into the wireless-application roadmaps [1.7], incessantly attempting to improve the quality of human life.

3. RESEARCH OBJECTIVES

This book deals with the design of analog-baseband architectures and circuits for *multistandard-compliant* and *low-voltage* wireless transceivers. As made apparent in the previous section, both design focuses are the key ingredients for a successful development of the emerging wireless products. Their implications to the silicon realization are briefly summarized below.

3.1 Multistandard-compliant analog-baseband architectures

As the concept of a software-defined radio has not yet been made mature enough for commercial reality, the current multistandard wireless transceiver solutions still call for dedicated RF front-ends for the bands of interest, such that the existed single-purpose solutions can be simply joined and reused. In view of that, as shown in Figure 1-3, a reconfigurable analog baseband becomes very helpful to interface the dedicated RF front-ends to the digital baseband that is able to be reconfigured by software.

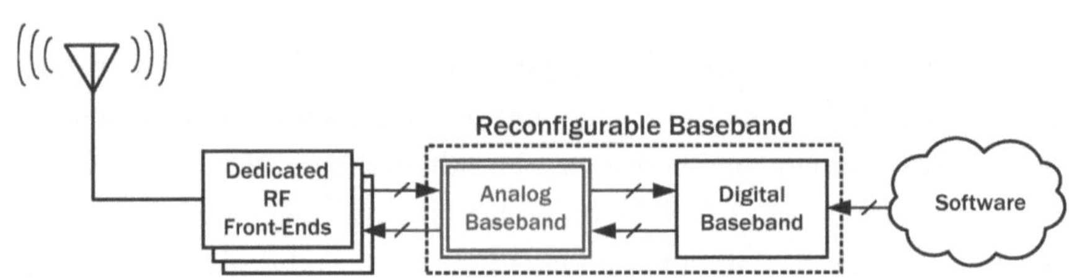

Figure 1-3. Design goal 1: a reconfigurable "analog baseband" for multistandard transceivers

The first design goal of this work is to develop reconfigurable analog-baseband architectures for an efficient multistandard compliance. The development starts from the architecture selection and frequency plan, to the design of functional blocks including IF mixer, channel-selection filter, programmable-gain amplifier (PGA) and dc-offset canceler (DOC). A/D and digital-to-analog (D/A) converters are beyond the topics of this book as they are assumed to be integrated inside the digital baseband to minimize the pin counts in the baseband interface.

3.2 Low-voltage analog-baseband functional blocks

Predicted by the International Technology Roadmap for Semiconductors (ITRS) [1.8], CMOS scaling leads to a continual shrink of supply voltage from micrometer to nanometer regimes (Figure 1-4(a) and (b)).

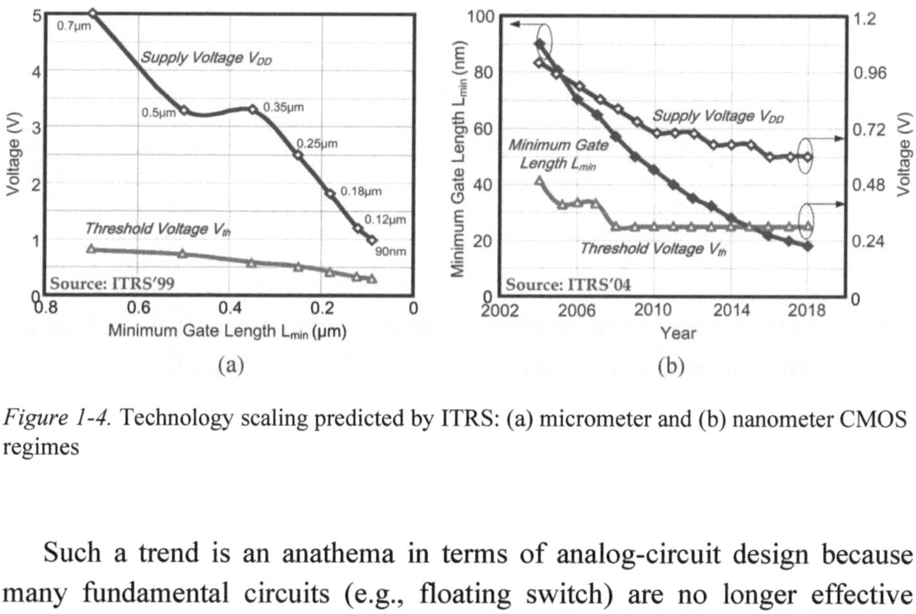

Figure 1-4. Technology scaling predicted by ITRS: (a) micrometer and (b) nanometer CMOS regimes

Such a trend is an anathema in terms of analog-circuit design because many fundamental circuits (e.g., floating switch) are no longer effective when the threshold voltage (V_{th}) reaches roughly 50% of the supply voltage (V_{DD}). The trade-off between performance and power becomes less straightforward, demanding low-voltage techniques to overcome the difficulties encountered in the realization.

As plotted in Figure 1-5, though low-voltage techniques have been extensively investigated to enable very low-voltage operation of continuous-time filters (CT-LPFs) [1.9][1.10][1.11][1.12] and sigma-delta (CT-ΣΔ)

modulators [1.13][1.14][1.15][1.16][1.17][1.18][1.19], additional varieties of functional blocks such as PGA, IF mixer and DOC, are also essential to form a (sub)system. In this work, the second design goal is to investigate more functional blocks that are able to operate at low voltage and high frequency (i.e., extend the state-of-the-art voltage-bandwidth boundary).

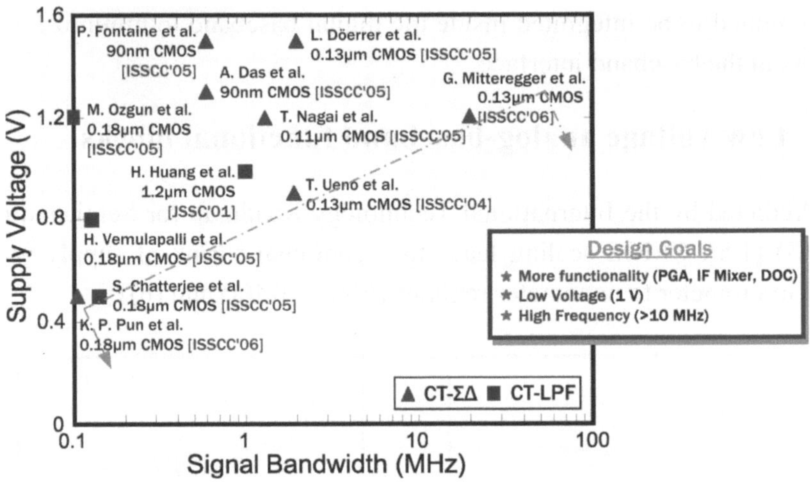

Figure 1-5. State-of-the-art low-voltage blocks (CT-ΣΔ and CT-LPF) and design goal 2: new low-voltage functional blocks that can operate at a high frequency

3.3 A fully integrated multistandard-compliant low-voltage analog-baseband platform for wideband applications

The final design goal is to demonstrate, simultaneously, the feasibility of the proposed multistandard architecture and low-voltage circuit techniques through the realization of a fully integrated analog-baseband platform for IEEE 802.11a/b/g-WLAN receivers. It involves the challenges of meeting all standards with minimal overhead and system reconfiguration, and operating power efficiently under a low-voltage supply of just 1 V.

REFERENCES

[1.1] Y. Neuvo, "Cellular Phones as Embedded Systems," *IEEE International Solid-State Circuits Conference (ISSCC), Digest of Technical Papers*, pp. 32–37, Feb. 2004.

[1.2] J. McNeill, M. Colin and B. Larivee, "A Split-ADC Architecture for Deterministic Digital Background Calibration of a 16b 1MS/s ADC," *IEEE International Solid-State Circuits Conference (ISSCC), Digest of Technical Papers*, pp. 276–277, Feb. 2005.

[1.3] G. K. Fedder and T. Mukherjee, "Tunable RF and Analog Circuits Using On-Chip MEMS Passive Components," *IEEE International Solid-State Circuits Conference (ISSCC), Digest of Technical Papers*, pp. 390–391, Feb. 2005.

[1.4] M.-A. Dubois, C. Billard, C. Muller, G. Parat and P. Vincent, "Integration of High-Q BAW Resonators and Filters Above IC," *IEEE International Solid-State Circuits Conference (ISSCC), Digest of Technical Papers*, pp. 392–393, Feb. 2005.

[1.5] B. Razavi, "A 60GHz Direct-Conversion CMOS Receiver," *IEEE International Solid-State Circuits Conference (ISSCC), Digest of Technical Papers*, pp. 400–401, Feb. 2005.

[1.6] P.-C. Huang, M.-D. Tsai, H. Wang and C.-H. Chen "A 114GHz VCO in 0.13μm CMOS Technology," *IEEE International Solid-State Circuits Conference (ISSCC), Digest of Technical Papers*, pp. 404–405, Feb. 2005.

[1.7] D. Chin, "Nanoelectronics for an Ubiquitous-Information society," *IEEE International Solid-State Circuits Conference (ISSCC), Digest of Technical Papers*, pp. 22–26, Feb. 2005.

[1.8] "The International Technology Roadmap for Semiconductors (ITRS) – *RF and Analog/Mixed-Signal Technologies for Wireless Communications*," 2004. [Online] Available: http://www.itrs.net/Common/2004Update/2004_04_Wireless.pdf

[1.9] H. Huang and E. K. F. Lee, "Design of Low-Voltage CMOS Continuous-Time Filter with On-Chip Automatic Tuning," *IEEE Journal of Solid-State Circuits (JSSC)*, vol. 36, no. 8, pp. 1168–1177, Aug. 2001.

[1.10] M. Ozgun, Y. Tsividis and G. Burra, "Dynamically Power-Optimized Channel-Select Filter for Zero-IF GSM," *IEEE International Solid-State Circuits Conference (ISSCC), Digest of Technical Papers*, pp. 504–505, Feb. 2005.

[1.11] S. Chatterjee, Y. Tsividis and P. Kinget, "A 0.5V Filter with PLL-Based Tuning in 0.18μm CMOS," *IEEE International Solid-State Circuits Conference (ISSCC), Digest of Technical Papers*, pp. 506–507, Feb. 2005.

[1.12] G. Vemulapalli, P. K. Hanumolu, Y.-J. Kook and U. K. Moon, "A 0.8-V Accurately Tuned Linear Continuous-Time Filter," *IEEE Journal of Solid-State Circuits (JSSC)*, vol. 40, no. 9, pp. 1972–1977, Sept. 2005.

[1.13] T. Ueno and T. Itakura, "A 0.9V 1.5mW Continuous-Time ΔΣ Modulator for WCDMA," *IEEE International Solid-State Circuits Conference (ISSCC), Digest of Technical Papers*, pp. 78–79, Feb. 2004.

[1.14] L. Dorrer, F. Kuttner, P. Greco and S. Derksen, "A 3mW 74dB SNR 2MHz CT ΔΣ ADC with a Tracking-ADC-Quantizer in 0.13μm CMOS," *IEEE International Solid-State Circuits Conference (ISSCC), Digest of Technical Papers*, pp. 492–493, Feb. 2005.

[1.15] T. Nagai, H. Satou, H. Yamazaki and Y. Watanabe, "A 1.2V 3.5mW ΔΣ Modulator with a Passive Current Summing Network and a Variable Gain Function," *IEEE International Solid-State Circuits Conference (ISSCC), Digest of Technical Papers*, pp. 494–495, Feb. 2005.

[1.16] P. Fontaine, A. N. Mohiedin and A. Bellaouar, "A Low-Noise Low-Voltage CT ΔΣ Modulator with Digital Compensation of Excess Loop Delay," *IEEE International Solid-State Circuits Conference (ISSCC), Digest of Technical Papers*, pp. 498–499, Feb. 2005.

[1.17] A. Das, R. Hezar, R. Byrd, G. Gornez and B. Haroun, "A 4th-Order 86dB CT ΔΣ ADC with Two Amplifiers in 90nm CMOS," *IEEE International Solid-State Circuits Conference (ISSCC), Digest of Technical Papers*, pp. 496–498, Feb. 2005.

[1.18] G. Mitteregger, C. Ebner, S. Mechnig, T. Blon, C. Holuigne, E. Romani, A. Melodia and V. Mellimi, "A 14b 20mW CMOS CT ΔΣ ADC with 20MHz Signal Bandwidth and 12b ENOB," *IEEE International Solid-State Circuits Conference (ISSCC), Digest of Technical Papers*, pp. 62–63, Feb. 2006.

[1.19] K.-P. Pun, S. Chatterjee and P. Kinget, "A 0.5V 74dB SNDR 25 kHz CT ΔΣ Modulator with Return-to-Open DAC," *IEEE International Solid-State Circuits Conference (ISSCC), Digest of Technical Papers*, pp. 72–73, Feb. 2006.

Chapter 2

TRANSCEIVER ARCHITECTURE SELECTION – REVIEW, STATE-OF-THE-ART SURVEY AND CASE STUDY

1. INTRODUCTION

Realizing multistandard transceivers with maximum hardware reuse amongst the given standards is of great importance to minimize the manufacturing cost of emerging multiservice wireless terminals. A well-defined architecture in conjunction with a reconfigurable building-block synthesis is essential to formulate such kind of tunable transceiver under a wide range of specifications. In this chapter, we present both fundamental and state-of-the-art techniques that help selecting transceiver architecture for single-/multi-standard design. It starts off by reviewing the basic schemes and examining their suitability for use in modern wireless communication systems (GSM, WCDMA, IEEE 802.11, Bluetooth, ZigBee and Ultra Wideband). The justifications are confirmed with the state-of-the-art choices via a survey (with 100+ references) of the most exploitation receiver and transmitter architecttures reported in 1997–2005 IEEE solid-state circuit forums: ISSCC, CICC, VLSI and ESSCIRC. State-of-the-art techniques for multi-standability are analyzed through careful case studies of a cellular receiver for GSM/DCS/PCS/WCDMA, and several WPAN/WLAN transceivers for Bluetooth/802.11b and IEEE 802.11a/b/g. They disclose, on the architecture and circuit levels, many ideas that have successfully inspired the recent development of wireless circuits and systems.

2. RECEIVER (RX) ARCHITECTURE

2.1 Superheterodyne receiver

The high reliability of superheterodyne RX [2.1.1] has made it the dominant choice for many decades. Its generic scheme is shown in Figure 2-1. With a band-selection filter rejecting the out-of-band interference, the in-band radio-frequency (RF) channels are free from amplification by a low-noise amplifier (LNA). A high-Q off-chip image-rejection filter prevents the image channel from being superimposed into the desired channel in the RF-to-intermediate frequency (IF) downconversion. The channel selection requires a voltage-controlled oscillator (VCO) driven by an RF frequency synthesizer and a high-Q off-chip surface-acoustic-wave (SAW) channel-selection filter (CSF). The signal level of the selected channel is properly adjusted by a programmable-gain amplifier (PGA) prior to the IF-to-baseband (BB) quadrature downconversion. This downconversion requires another phase-locked loop (PLL) and a quadrature VCO (QVCO) for generating the in-phase (I) and quadrature-phase (Q) components. The BB lowpass filters (LPFs) are of low-Q requirement but high in filter order for ultimate channel selection. The BB PGAs adjust the signal swing for an optimum-scale analog-to-digital (A/D) conversion.

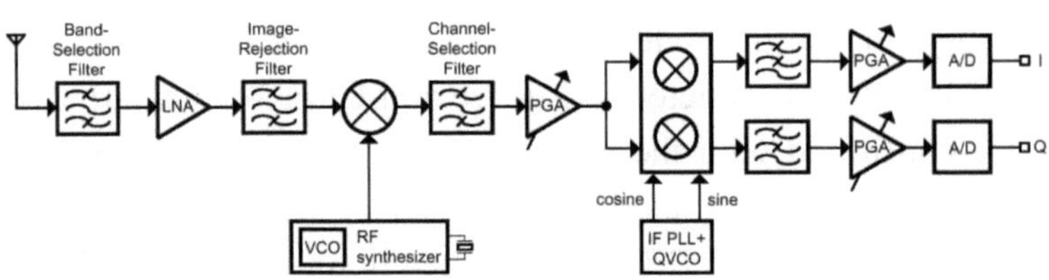

Figure 2-1. Superheterodyne RX

The superior I/Q matching because of low operating frequency, as well as the avoidance of dc-offset and $1/f$-noise problems, are the pros of superheterodyne. Whereas the low integration level and high power consumption for the on/off-chip buffering are its cons. There also exists a trade-off in IF selection; a high IF (e.g., ~70 MHz) improves the sensitivity due to higher attenuation can be offered by the image-rejection filter, while a low IF (e.g., ~10 MHz) enhances the selectivity due to a lower Q requirement from the SAW filter. On the other hand, due to the restricted IF choice of 10.7 or 71 MHz for commercial filters, a superheterodyne RX for multistandard design normally constitutes a high cost for filtering at different IFs.

2.2 Image-rejection receiver – Hartley and Weaver

The principle of image-rejection RX is to process the desired channel and its image in such a way that the image can be eliminated eventually by *signal cancellation*. Hartley [2.1.2] proposed such an idea with the architecture shown in Figure 2-2(a). The RF signal in the downconversion is split into two components by using two matched mixers, a QVCO and an RF synthesizer. The outputs, namely in phase (I) and quadrature phase (Q), are then filtered by the LPFs. With a 90°-phase-shifter added to the Q channel, the image can be canceled after the summation of I and Q outputs. In practice, a perfect-quadrature downconversion and a precise 90°-phase-shifter cannot be implemented in analog domain, especially at high frequency. The practical values of static gain/phase mismatches are 0.2–0.6 dB/1–5°, corresponding to an image rejection of roughly 30–40 dB.

The Weaver RX [2.1.3], as shown in Figure 2-2(b), features a similar principle as Hartley's one, except the 90°-phase-shifter that is replaced by a quadrature downconverter. The key advantage of such a replacement is related with the fact that a downconverter can realize relatively a much wideband quadrature matching. The overheads are the additional IF mixers, PLL and QVCO. Both Hartley and Weaver schemes feature high integratability and are convenient to use in multistandard design.

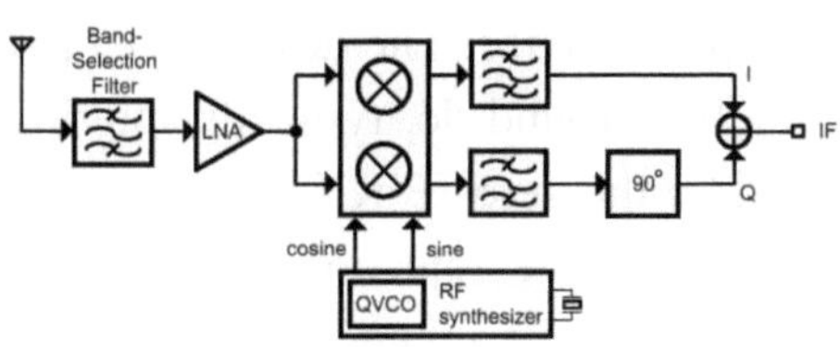

(a)

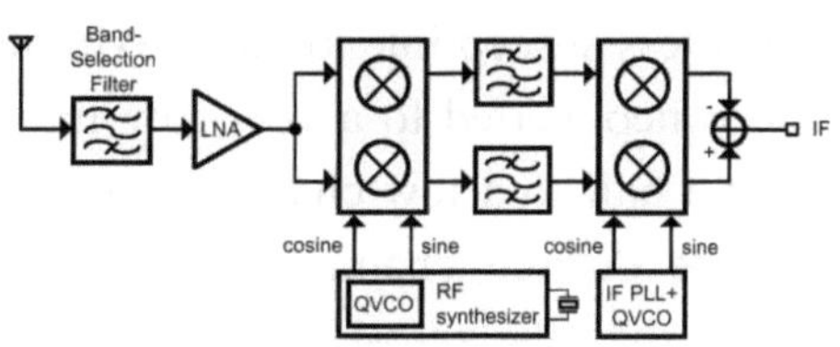

(b)

Figure 2-2. Image-rejection RX: (a) Hartley and (b) Weaver

2.3 Zero-IF receiver

Similar to image-rejection RX, a zero-IF RX [2.1.4] obviates using off-chip filters. As shown in Figure 2-3, the desired channel is translated directly to dc through two channels (I and Q) that operate with a 90°-phase shift. Again, the image is eliminated through signal cancellation rather than filtering. Since the image is the desired channel itself, the demanded I/Q matching is practically achievable for most applications.

The fundamental limitation of zero-IF RX is its high susceptibility to low-frequency interference, i.e., dc offset and $1/f$ noise. With them superimposed on the desired channel, a substantial degradation in signal-to-noise ratio (SNR) or completely desensitizing the system due to a large baseband gain may result. A capacitive coupling and a servo loop are common choices to alleviate this problem, but at the expense of long settling time and large chip area for realizing the very low cutoff-frequency highpass pole.

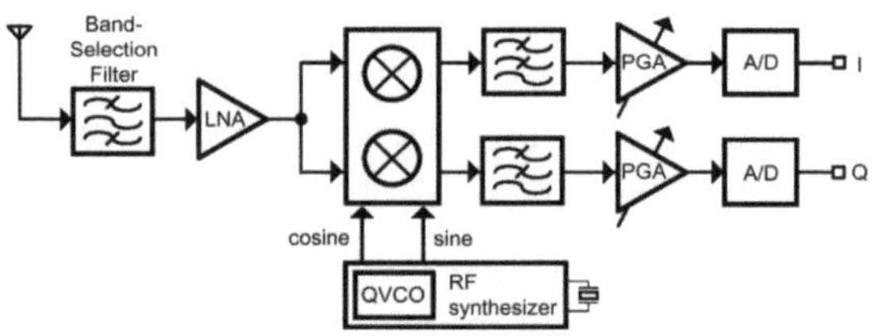

Figure 2-3. Zero-IF RX

2.4 Low-IF receiver

Low-IF RX [2.1.5] features a similar integratability as the zero-IF one, but is less susceptible to the low-frequency interference. The desired channel is downconverted to a very low-frequency bin around dc, typically ranging from a half to a few channel spacings. Unlike the zero-IF RX, the image is not the desired channel itself. The required image rejection is normally higher than the zero-IF approach as the power of the image can be significantly larger than that of the desired channel. Depending on the permutation of the building blocks, a low-IF RX can be implemented in many ways. Four examples are shown in Figure 2-4(a)–(d).

Case-I performs the IF-to-BB downconversion digitally, eliminating the secondary image problem while permitting a pole-frequency-relaxed dc-offset cancellation adopted in the analog BB. The disadvantages are a higher bandwidth requirement (compared with the zero-IF) from the LPFs and PGAs, and a higher conversion-rate requirement from the A/Ds.

Case-II is identical to Case-I except the LPFs are replaced by a pair of filters operating in complex domain, namely a polyphase filter or a complex filter. Such a filter performs not only channel selection, but also relaxes the I/Q-matching requirement from the PGAs and A/Ds.

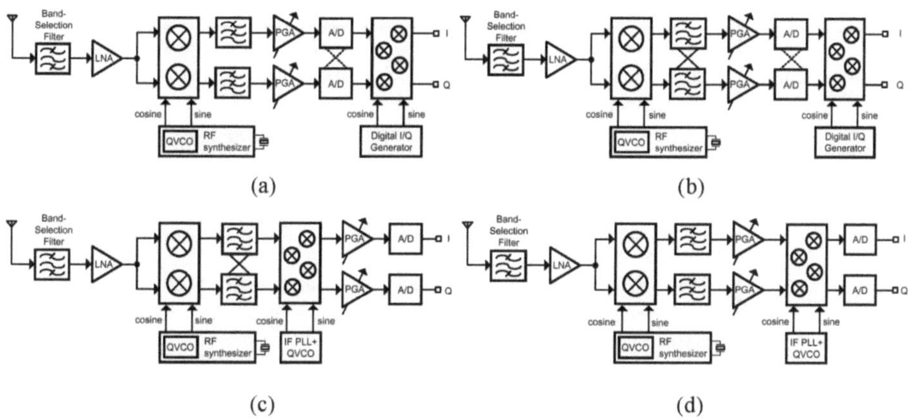

Figure 2-4. Low-IF RX implemented in different ways: (a) case-I (b) II (c) III and (d) IV

Case-III is another combination employing a complex filter together with an analog IF-to-BB downconverter for doubling the image rejection. With such a structure, the I/Q matching required from the following PGA and A/D are very relaxed. The bandwidth of the PGAs and the conversion rate of the A/Ds are reduced to their minimum like zero-IF. The associated overhead is a low cutoff highpass pole in the dc-offset cancellation that is necessary in the PGAs due to a high baseband gain. The chip-area impact is therefore very significant since the systems containing I and Q channels are typically differential (i.e., four highpass poles).

Case-IV positions an analog IF-to-BB downconverter prior to the A/Ds such that the conversion rate of the A/Ds can be minimized. Different from

Case-III, the dc-offset cancellation for the PGA and LPF is relaxed in its highpass pole frequency.

2.5 Comparison of different receiver architectures

Table 2-1 summarizes and compares the presented RX architectures. Their characteristics determine their appropriateness for use in modern wireless communication systems, as presented in *Section 4*.

Table 2-1. Summary of different RX architectures

RX Architecture	Advantages	Disadvantages
Superheterodyne	+ Reliable performance + Flexible frequency plan + No DC offset and 1/*f* noise	– Expensive and bulky, high power – Difficult to share the SAW filters for multistandard
Image-Rejection (Hartley and Weaver)	+ Low cost + No DC offset and 1/*f* noise + High integratability	– Quadrature RF-to-BB downconversion – Suffer from first and secondary images – Narrowband (Hartley) – High I/Q matching
Zero-IF	+ Low cost + Simple frequency plan for multistandard + High integratability + No Image problem	– Quadrature RF-to-BB downconversion – DC-offset and 1/*f*-noise problems
Low-IF	+ Low cost + High integratability + Small DC offset and 1/*f* noise	– Image is a problem – Quadrature RF-to-IF and double-quadrature IF-to-BB downconversions

3. TRANSMITTER (TX) ARCHITECTURE

3.1 Superheterodyne transmitter

Architecturally, the superheterodyne TX (Figure 2-5) is a reverse of operation from its RX counterpart with the A/D conversion replaced by a digital-to-analog (D/A) conversion. However, they are very different in design specification. For instance, in transmission, only one channel will be upconverted in the TX. Its power level is well determined throughout the TX path. Differently in signal reception, the power of the incoming signals is variable and the desired channel is surrounded with numerous unknown-power in-band and out-of-band interferences. Thus, PGAs are essential for the RX to relax the dynamic range of the A/D converter, but may be omitted in the TX if the power control could be fully implemented by the power amplifier (PA). Similarly, since the channel in the TX is progressively amplified toward the antenna and finally radiated by a PA, the linearity of the whole TX is dominated by the PA. Whereas it is the noise contribution of the LNA dominates the whole RX noise figure.

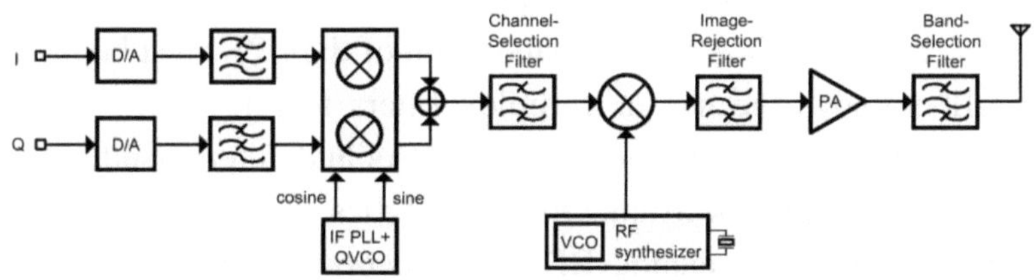

Figure 2-5. Superheterodyne TX

3.2 Direct-up transmitter

The direct-up TX (Figure 2-6) features an equal integratability as the zero-IF RX, but is limited by the well-known LO pulling. To meet the standard required modulation mask, techniques such as offset VCO and LO-leakage calibration are somehow necessary. Again, it is noteworthy that albeit the functional blocks in RX and TX are identical, their design specifications are largely different. For instance, the RX-LPF has to feature a high out-of-band linearity due to the coexistence of adjacent channels, whereas it is not demanded from TX-LPF.

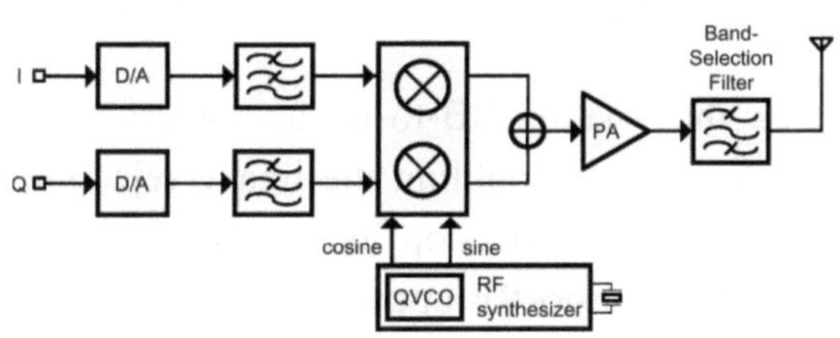

Figure 2-6. Direct-up TX

3.3 Two-step-up transmitter

Similar to the low-IF RX, two-step-up TXs can be structured into four possible schemes as shown in Figure 2-7(a)–(d).

Case-I frequency-up-translates the desired channel digitally prior to D/A conversion such that the low-frequency interference from the D/A and LPF can be canceled by means of an ac-coupling or a servo loop, avoiding the transmission of DC-to-LO-mixing products. In this case, the required conversion rate of the D/A and the bandwidth of the LPF must be increased.

Case-II employs the use of complex filters to reject the image raised from I/Q mismatch between the I and Q D/As and filters itself to improve the purity of the output spectrum. Other properties are similar to Case-I.

Case-III employs an analog BB-to-IF upconverter to reject the image. With such permutation, only a LPF would be required and the output spectrum is purified.

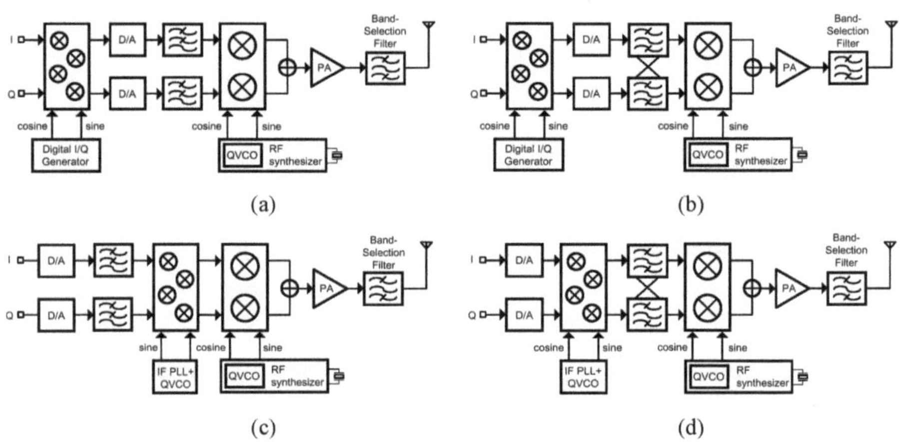

Figure 2-7. Two-step-up TX implemented in different ways: (a) case-I (b) II (c) III and (d) IV

Case-IV locates the analog BB-to-IF upconverter between the D/A converter and complex filters, doubling the image rejection and allowing a capacitive coupling (or by a servo loop) between the upconverter and filter, and between the filter and IF-to-RF upconverter. One key advantage of this scheme is the allowance of independent dc-biasing for each block.

Compared with the direct-up TX, the LO feedthrough is reduced (of course, the amount depends on the selected IF and port-to-port isolation) as the first and second VCOs can be offset from each other (i.e., LO=VCO_1+ VCO_2). The overheads are the additional power and area consumption required for the mixing, filtering and frequency synthesis.

3.4 Comparison of different transmitter architectures

Table 2-2 summarizes and compares the presented TX architectures. As it will be presented in the next section, their characteristics determine their appropriateness for use in modern wireless communication systems.

Table 2-2. Summary of different TX architectures

TX Architecture	Advantages	Disadvantages
Superheterodyne	+ Reliable performance + Flexible frequency plan + No LO leakage + Simple DC-offset cancellation at BB	– Expensive and bulky, high power – Difficult to share the SAW filters for multistandard
Direct-Up	+ Low cost + Simple frequency plan for multistandard + High integratability + No Image problem	– Quadrature BB-to-RF downconversion – LO leakage – DC-offset cancellation is difficult at BB (area and settling time impacts)
Two-Step-up	+ Low cost + High integratability + Simple DC-offset cancellation at BB	– Image is a problem – Quadrature IF-to-RF and double-quadrature BB-to-IF downconversions – LO leakage (depends on the IF)

4. RX AND TX ARCHITECTURES FOR MODERN WIRELESS COMMUNICATION SYSTEMS

4.1 GSM/DCS/PCS

GSM [2.1.6] and its copies: DCS and PCS are currently the most dominated standards for cellular communication. Except their differences in frequency band and geographical use, the PHY-relevant data are alike, as listed in Table 2-3. With the Gaussian minimum shift keying (GMSK) as the modulation method; a time-division 200-kHz channel can deliver a data rate of 270 kb/s. RXs designed for GSM/DCS/PCS using superheterodyne [2.2.1], low-IF [2.2.2] or zero-IF [2.2.6] architecture have all been tried.

The TX typically follows the RX in architecture to share the SAW filter (if any) and frequency synthesizing components. Examples are [2.2.1] using superheterodyne RX with TX and [2.2.2] using a zero-IF/low-IF RX with a direct-up TX. In addition to them, there are other possible types of TX. The two-step-up TX [2.2.16] can gain advantages from the frequency relationship of GSM (0.9 GHz) and DCS (1.8 GHz) to realize a dual-mode solution. Additionally, PLL [2.2.18] and polar [2.2.19] modulation-based TXs are also possible due to the constant amplitude of GMSK signal.

Table 2-3. GSM/DCS/PCS characteristics

	GSM	DCS	PCS
Modulation	GMSK	GMSK	GMSK
Frequency Band	890–960 MHz	1,710–1,850 MHz	1,880–1,930 MHz
Channel Bandwidth	200 kHz	200 kHz	200 kHz
Bit Rate	270 kb/s	270 kb/s	270 kb/s

4.2 WCDMA (UMTS)

The 3G wireless system is known as WCDMA or UMTS [2.1.7]. It can deliver a data rate up to 3.84 Mb/s. A pseudorandom sequence spreads the quadrature phase-shift keying (QPSK) modulated signal to a 3.84-MHz bandwidth. Its main characteristics are listed in Table 2-4.

The wideband and spread spectrum nature of WCDMA stimulates the use of zero-IF RX [2.2.25], providing that the dc offset and $1/f$ noise are comfortably eliminated through, for instance, ac coupling or servo loop. The required image rejection with a zero-IF is very relaxed, i.e., ~25 dB. The induced intersymbol interference (ISI) is uncritical as long as the highpass pole of the dc-offset cancellation is sufficiently small (<10 kHz). The concurrent transmit and receive operations, however, require a very linear RF path.

WCDMA TXs using superheterodyne [2.2.35], direct-up [2.2.36] or two-step-up [2.2.37] architecture have all been tried. However, if a zero-IF RX has been chosen, a direct-up TX is efficient to follow such that the blocks for frequency synthesis can be reused. The problem of LO pulling requires calibration circuits (or other type of circuitry) to suppress the carrier leakage [2.2.36].

Table 2-4. WCDMA characteristics

	WCDMA
Modulation	QPSK
Frequency Band	1,920–2,170 MHz
Channel Bandwidth	3.84 MHz
Bit Rate	3.84 Mb/s

4.3 802.11x and HiperLAN 2

WLAN is intended to provide high-speed internet access whenever wired LANs are not possible (also economically feasible), or many subscribers are diverged within a relatively large place such as airport or hotel. The IEEE 802.11 [2.1.8] family is dedicated for high-speed WLAN communication. Currently, the most relevant PHY-layers are 802.11a, b and g. The older FH and DS modes are seldom used today, while the latest 802.11n will be ratified soon.

The 802.11FH and 802.11DS operating in the 2.4-GHz ISM band provide a maximum data rate of 2 Mb/s by using the frequency-hopping spread spectrum (FHSS) and direct-sequence spread spectrum (DSSS) techniques, respectively. The 802.11a based on the orthogonal frequency-division multiplexing (OFDM) technique delivers a high data rate up to 54 Mb/s by using 64-quadrature amplitude modulation (64-QAM) in the 5-GHz UNII band. For 802.11b, the maximum data rate is 11 Mb/s by exploiting Complementary Code Keying (CCK) modulation in the 2.4-GHz band. A mix of a and b is g, which supports a data rate up to 54 Mb/s using the OFDM with 64-QAM in the 2.4-GHz ISM band, and backward complies with b for lower data-rate options. The HiperLAN 2 [2.1.9] is harmonized with the 802.11a. Their characteristics are tabulated in Table 2-5.

The wideband nature of 802.11a/b/g and HiperLAN again suggests the use of zero-IF RX [2.2.39]. However, the low-IF [2.2.42] is found effective for OFDM mode in 802.11a/g, a frequency error in frequency conversion can locate the ±1st subcarriers on the notch of the dc-offset cancellation. The dual-conversion [2.2.45] is similar to superheterodyne but using a relatively high IF (such as 3 GHz) to allow the image rejection to be fully realized on-chip. Moreover, since the channel selection is at IF, lower power and better phase-noise LO is expected when comparing it with the zero-IF design. The frequency plan is critical to maximize the functional blocks sharing in the dual-conversion.

The architectural choice of WLAN TX is typically consistent with that of the RX.

Table 2-5. 802.11x and HiperLAN 2 characteristics

	802.11FH	802.11DS	802.11a	802.11b	802.11g	HiperLAN 2
Modulation	FHSS: D-BPSK/ D-QPSK	DSSS: D-BPSK/ D-QPSK	OFDM: BPSK/ QPSK/ QAM	DSSS: D-BPSK/ D-QPSK CCK	DSSS: D-BPSK/D-QPSK CCK OFDM: BPSK/ QPSK/QAM	OFDM: FSK/GMSK
Frequency Band	2.4 GHz [a]	2.4 GHz [a]	5 GHz [b]	2.4 GHz [a]	2.4 GHz [a]	5 GHz [c]
Channel Bandwidth	1 MHz	22 MHz	16.25 MHz	22 MHz	16.25–22 MHz	16.25 MHz
Bit Rate	1, 2 Mb/s	1, 2 Mb/s	6–54 Mb/s	1–11 Mb/s	1–54 Mb/s	1.5–54 Mb/s

Remark: a: 2,402–2,480 MHz, b: 5,150–5,350 and 5,725–5,825 MHz, c: 5,150–5,350 and 5,470–5,725 MHz

4.4 Bluetooth (802.15.1), HomeRF, ZigBee (802.15.4) and Ultra Wideband (802.15.3)

A couple of standards have been deployed for wireless personal-area network (WPAN) focusing on low-cost low-power RF technology. The dominant standards are Bluetooth, HomeRF, ZigBee and Ultra Wideband (UWB). Their characteristics are tabulated in Table 2-6.

Bluetooth, also known as 802.15.1 [2.1.10], uses FHSS with Gaussian frequency shift keying (GFSK) as the modulation, delivering a data rate of 1 Mb/s using a 1-MHz channel at the 2.4-GHz ISM band. Most of the solutions for Bluetooth are low-IF RXs with direct-up TXs [2.2.58] for their simplicity but adequate performances. The PLL TXs [2.2.67][2.2.68][2.2.69] [2.2.70] are also proved to be feasible since the frequency synthesizer can generate a constant-amplitude GFSK channel without a separated TX path.

HomeRF [2.1.11] is similar to Bluetooth; it also uses FHSS access technique, GFSK modulation and the 2.4-GHz ISM band, being the exception the channel bandwidth that is 5 MHz for a higher data rate of 5 Mb/s. Many characteristics of HomeRF align with Bluetooth. With a bandwidth-tunable baseband, a Bluetooth solution can be reused for HomeRF.

ZigBee, also known as 802.15.4 [2.1.12], is another low-cost wireless technology. It employs a 5-MHz GFSK channel to communicate at the 2.4-GHz ISM band, delivering a data rate up to 250 kb/s. A complete TXR has been reported for this standard [2.2.78]. Consistent to our previous considerations, its narrowband nature recommends the use of a low-IF RX together with a direct-up TX.

UWB, also known as 802.15.3 [2.1.13], is an evolutionary standard highly different from the existing ones. One UWB band is between 3.1 GHz and 10.6 GHz with a bandwidth of 500 MHz, delivering a data rate up to 480 Mb/s using shaped-pulse (SP) or frequency-switched OFDM (FS-OFDM) modulation. Such a wideband system suggests, again, the use of zero-IF RX [2.2.81] [2.2.82] and direct-up TX [2.2.83].

Table 2-6. Bluetooth, HomeRF, ZigBee and Ultra Wideband characteristics

	Bluetooth	HomeRF	ZigBee	Ultra Wideband
Modulation	FHSS: GFSK	FHSS: GFSK	GFSK	SP/FS-OFDM
Frequency Band	2,402–2,480 MHz	2,402–2,480 MHz	2,402–2,480 MHz	3,100–1,060 MHz
Channel Bandwidth	1 MHz	1 MHz, 5 MHz	5 MHz	500 MHz
Bit Rate	1 Mb/s	1.6–10 Mb/s	20, 40, 250 kb/s	110–480 Mb/s

5. SURVEY OF THE STATE-OF-THE-ART WORKS FOR MODERN WIRELESS STANDARDS

The research and development on wireless systems have turned what seemed impractical systems like zero-IF RX to plausible solutions. Simultaneously, the conventional superheterodyne architecture has started to fade out due to its high power and high cost. This section will present statistic results of RX and TX architectures that have been employed in the state-of-the-art works. The papers are collected from the IEEE solid-state circuit forms: ISSCC, VLSI, CICC and ESSCIRC from 1997 to 2005. With the abbreviation of each type RX and TX architecture listed in Table 2-7, the distributions of each type of RX and TX employed in the modern wireless communication systems are plotted in Figure 2-8 and 2-9, respectively. The publications are listed in the References Part II. They are classified by their applications and corresponding architectures.

Table 2-7. Abbreviations of TX and RX architectures

SH: Superheterodyne	LIF: Low-IF
PLL: Phased-locked Loop Modulation	ZIF: Zero-IF
POLAR: Polar Modulation	DU: Direct-up
DUAL: Dual-Conversion	2SU: Two-Step-Up
WEAVER: Weaver	HIF: High IF

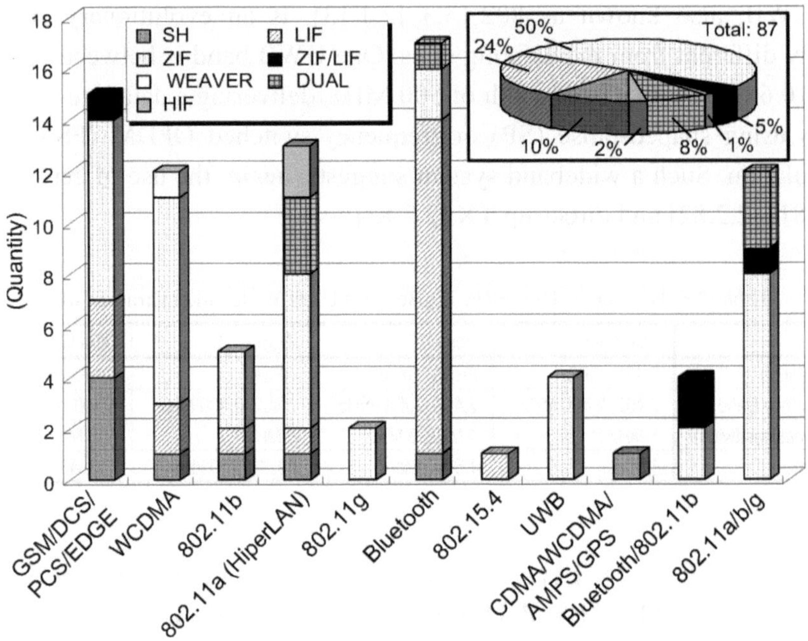

Figure 2-8. RX architectures employed in the state-of-the-art works

For the RX, it is obvious that the dominance of superheterodyne RX has been replaced by the zero-IF and low-IF solutions. The zero-IF RX has been tried for all applications except the 802.15.4 and the CDMA/WCDMA/AMPS/GPS. Whereas the low-IF RX has been widely used for narrowband applications such as Bluetooth, GSM/DCS/PCS and 802.15.4. Dual-conversion has also been proved effective for high-frequency standards like the 802.11a and HiperLAN.

For the TX, the dominant choice is direct-up. It is not only because of its integratability and low power, but also to match the RX path that uses zero-IF. The same consideration is applicable for two-step-up and dual-conversion. PLL TX has been tried for GSM/DCS/PCS and Bluetooth due to the constant amplitude of their modulation signals. A direct modulation during frequency synthesis eliminates the need of a TX path.

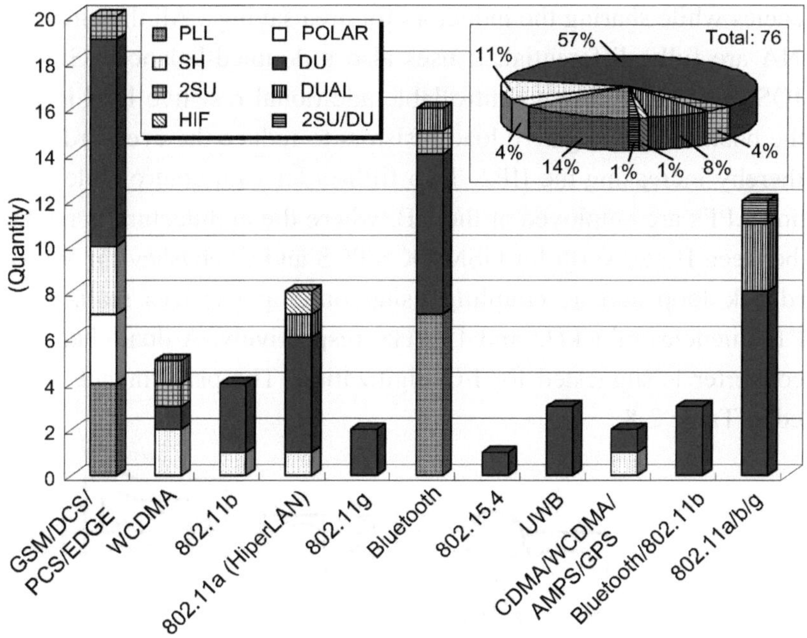

Figure 2-9. TX architectures employed in the state-of-the-art works

The following subsections discuss the state-of-the-art RXs and transceivers (TXRs) designed for cellular, WPAN/WLAN and WLAN applications, and compare their achieved performances in relation with their architectural choice and levels of block sharing.

6. CASE STUDY

A multistandard transceiver should satisfy the given standards with the minimum increase in silicon area and complexity. Reconfigurability would then be a mandatory strategy to be adopted for reaching such goals. This section discusses the key techniques applied in the state-of-the-art multistandard wireless receivers and transceivers.

6.1 Cellular receiver: GSM/DCS/PCS/WCDMA

J. Ryynänen et al. [2.1.14] – it is a direct-conversion RX (Figure 2-10) designed for GSM/DCS/PCS/WCDMA. It contains a singled-ended bipolar LNA employing a parallel transistor to select the gain peak at different frequencies while sharing the inductors for area savings. All the circuits after the LNA are fully differential. It uses also a doubled-balanced Gilbert-cell BiCMOS mixer featuring a controllable additional resistive load in parallel with the positive and negative load resistors to reduce the even-order distortion, thereby increasing the IIP2. Two fifth-order gain-controllable channel-selection LPFs are employed at the BB, where the architecture is reconfigurable between Butterworth for GSM/DCS/PCS and Chebyshev for WCDMA. A feedback loop and ac coupling using on-chip passives realizes –3-dB cutoff frequencies of 1 kHz and 13 kHz, respectively. A dual-channel 8-bit A/D converter is suggested for I/Q digitization. The performance summary is listed in Table 2-8.

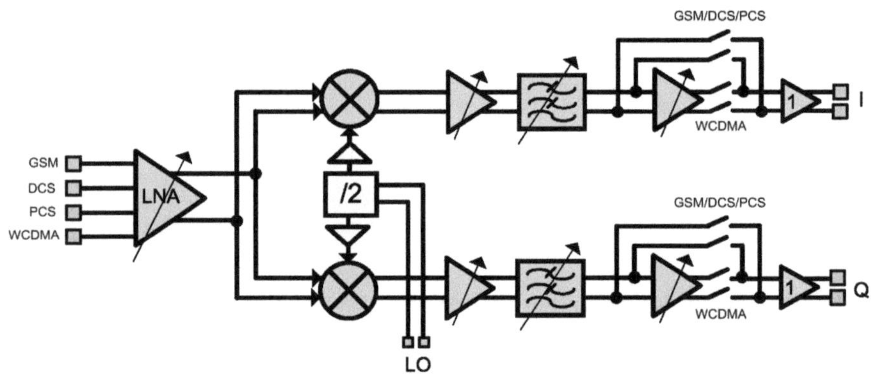

Figure 2-10. J. Ryynänen's GSM/DCS/PCS/WCDMA receiver

Table 2-8. Performance summary of J. Ryynänen's GSM/DCS/PCS/WCDMA receiver

	GSM/DCS/PCS	WCDMA
RX Architecture	Zero-IF	
Integration	Receiver (no A/D, synth. and VCO)	
Technology	0.35-μm SiGe BiCMOS	
Supply	2.7 V	
Chip Area	9.8 mm^2	
Noise Figure	2.8 to 4.8 dB	3.5 dB
IIP3	–20 to –21dB	–21 dB
IIP2	+42 dB	+47 dB
Power	42 mW	50 mW

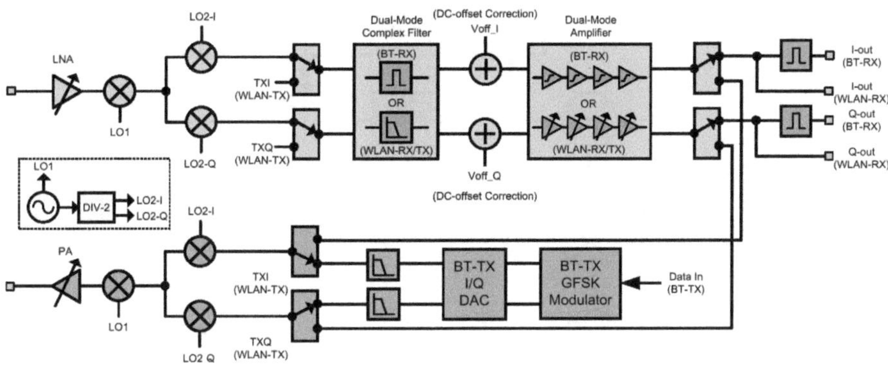

Figure 2-11. T. Cho's Bluetooth/802.11b transceiver

6.2 WPAN/WLAN transceivers for Bluetooth/802.11b

T. Cho et al. [2.2.87] – this work employs a two-step-down/up TXR architecture (Figure 2-11) to reduce the LO-to-LO self-mixing. The first IF is one-third of the RF to simplify the generation of the second LO (dividing by 2 the value of the first LO), whereas the second IF is 2 MHz/0 Hz for Bluetooth /802.11b. Because Bluetooth and 802.11b share the identical 2.4 GHz ISM band, the LNA, first and second down/upconverters can all be shared. The channel-selection filter is a fifth-order Butterworth 1-MHz complex filter centered at 2-MHz IF for Bluetooth, whereas it is reconfigured as a 7.5-MHz LPF centered at dc for 802.11b. For the baseband amplification, it is a limiting/automatic gain control (AGC) amplifier for Bluetooth/ 802.11b. Both the filter and amplifier are shared between the RX and TX to further save silicon area. A wideband fractional-*N* frequency synthesizer satisfies the diverse step size of either mode and achieves a fast RX–TX turn-around time. DC-offset correction conducts in every frame using a D/A feedback.

Y. Jung et al. [2.2.88] – this direct-conversion TXR (Figure 2-12) shares all building blocks between Bluetooth and 802.11b modes. The sixth-order Butterworth LPFs have 0.7-/5.5-MHz bandwidth for Bluetooth/802.11b. The PGAs are embedded with dc-offset cancellation loops, which realize a 1/100 kHz –3-dB cutoff for Bluetooth/802.11b. A fractional-*N* frequency synthesizer with charge-averaging charge pump scheme for canceling low-frequency spur is used to satisfy several step sizes.

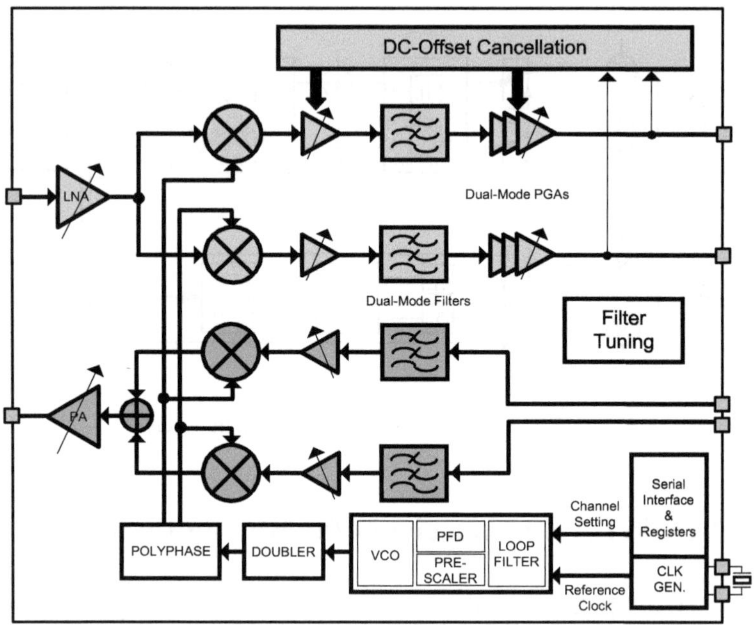

Figure 2-12. Y. Jung's Bluetooth/802.11b transceiver

H. Darabi et al. [2.2.86] – this TXR, as shown in Figure 2-13, features a low-IF/zero-IF RX for Bluetooth/802.11b and a direct-up TX for both. A 1.6-GHz VCO self-mixes with its division-by-2 product, i.e., 0.8-GHz, to generate the 2.4-GHz LO. The LNA, first and second down/upconverters are shared as both Bluetooth and 802.11b operate in the 2.4-GHz ISM band. Two channel-selection filters are implemented, one is a fifth-order 1-MHz complex filter centered at 2-MHz IF for Bluetooth, another is a fifth-order 7.5-MHz LPF centered at dc for 802.11b. For the baseband amplification, it is a limiter for Bluetooth and a PGA for 802.11b. A fractional-*N* frequency synthesizer satisfies diverse step sizes. Finally, the dc offset is removed by a programmable offset cancellation loop, which refreshes the control of the –3-dB cutoff frequency in every packet.

Discussion – all examples are complied with the standards under the general conclusion that low-IF and zero-IF approaches are suitable for narrow-band and wideband applications, respectively. As compared in Table 2-9 with the performances of the discussed three TXRs, the key advantage of T. Choi's TXR is its compact size due to the sharing not only of the RF front-end, but also the BB blocks between Bluetooth and 802.11b through reconfiguration,

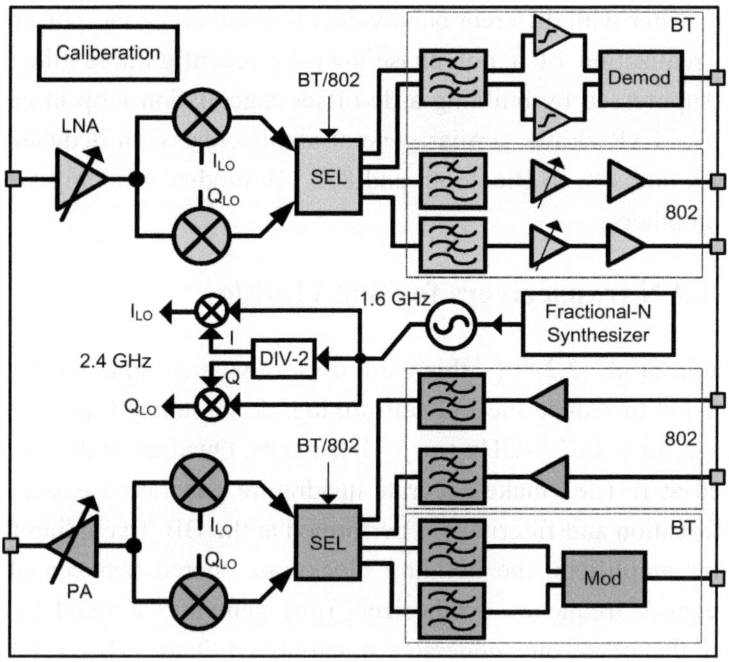

Figure 2-13. H. Darabi's Bluetooth/802.11b transceiver

Table 2-9. Performance comparison of Bluetooth/802.11b transceivers

	T. Cho et al.		Y. Jung et al.		H. Darabi et al.	
	Bluetooth	802.11b	Bluetooth	802.11b	Bluetooth	802.11b
RX Architecture	Dual-Conversion with Low-IF	Dual-Conversion with Zero-IF	Zero-IF		Low-IF	Zero-IF
TX Architecture	Direct-Up	Direct-Up	Direct-Up		Direct-Up	Direct-Up
Integration	Transceiver *		Transceiver *		Transceiver *	
Technology	0.18-μm CMOS		0.25-μm CMOS		0.35-μm CMOS	
Supply	1.8 V		2.7 V		3.0 V	
Chip Area	16 mm^2		8.4 mm^2		25 mm^2	
Sensitivity	80 dBm		–87 dBm	–86 dBm	–80 dBm	> –93 dBm
IIP3	–12/+14 dBm #		–15 dBm		–7 dBm +	–8 dBm +
Power in RX	108 mW		135 mW	175.5 mW	138 mW	195 mW
Power in TX	72 mW	126 mW	121.5 mW	162 mW	141 mW	234 mW

#: LNA high/low gain, +: minimum gain, *:no A/D and D/A for 802.11b

and in receive and transmit modes. However, the TX–RX turn-around time becomes an issue by permitting the use of time-continuous dc-offset cancellation loop, which results in a long settling time. Y. Jung's TXR achieves a very small area by using a direct-conversion for both RX and TX.

A lowpass filter with different bandwidths is much area- and power-efficient than the realization of a polyphase/lowpass reconfigurable filter. The dc offset is suppressed by building a dc-offset cancellation loop in each PGA. H. Darabi's TXR shows similar general architecture with dedicated analog BB for RX and TX, in Bluetooth and 802.11b modes. The chip area is thus larger than others.

6.3 WLAN transceivers for 802.11a/b/g

R. Ahola et al. [2.2.90] – this work uses a two-step-up/down architecture (Figure 2-14) to reduce the LO self-mixing. A common IF of 1.3–1.5 GHz was chosen for both 2.4-GHz and 5-GHz bands. Quadrature demodulation is performed at IF (i.e., make accurate quadrature generation easier) whereas the amplification and filtering are performed at the BB. Excluding the LNAs and power amplifiers, the building blocks are shared between all modes. Two integer-N frequency synthesizers (one generates a fixed LO at 3.84 GHz and the other one generates a variable LO at 1.3–1.5 GHz) were employed. The channel-selection LPFs in the receiver and transmitter paths are fourth-order Butterworth and fourth-order Chebyshev, respectively. Their bandwidths are accurately set by a calibration engine. Two highpass poles are inserted in the receiver path; one is in front of the variable-gain amplifier (VGA), and the other is a servo loop round the LPF. The former and latter pole frequencies are <200 Hz and <1 kHz in all process corners, respectively, ensuring that multipath fading and worst-case frequency offset result no significant performance degradation. The VGA's gain-tuning resulted dynamic dc offset is, in first order, minimized by setting the bias conditions of the amplifying devices constant in gain change. The measured dc-offset step in gain change never exceeds 10 mV with this technique.

Z. Xu et al. [2.2.93] – it is a direct-conversion TXR (Figure 2-15) for all 802.11a, b and g. Two LNAs for 2.4-GHz and 5-GHz bands were implemented with shared mixers, channel-selection filters and VGAs. One frequency synthesizer and one VCO shares between different modes, RX and TX. The LO leakage in the transmitter is eliminated by applying a LO-leakage calibration loop. The lowpass channel-selection filter in both RX and TX is of fifth-order and offers a variable gain feature. A successive switching technique implemented in the dc-offset cancellation loop is to improve the settling time in the RX gain adjustment.

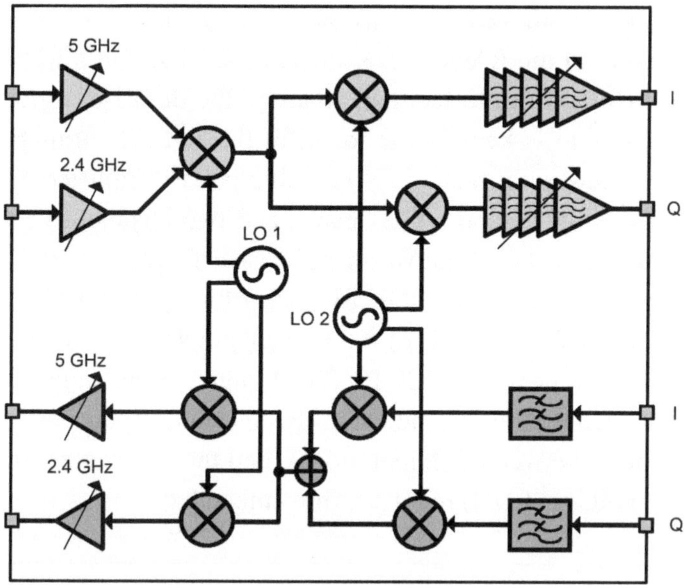

Figure 2-14. R. Ahola's 802.11a/b/g transceiver

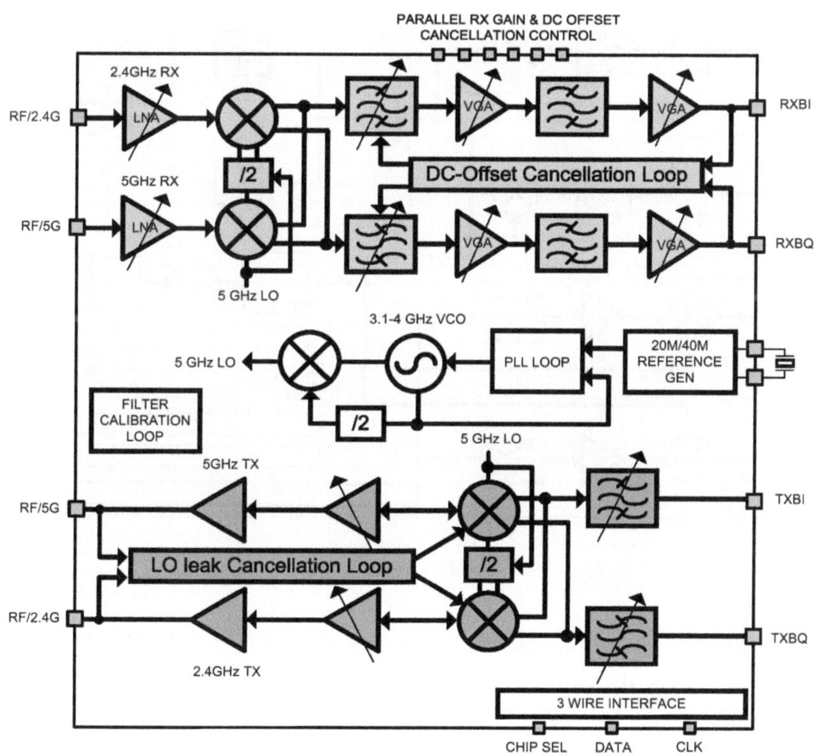

Figure 2-15. Z. Xu 802.11a/b/g transceiver

T. Rühlicke et al. [2.2.99] – as shown in Figure 2-16, it employs a direct-conversion in the RX for CCK mode (802.11b), but a low-IF one for OFDM mode (802.11a/g) to facilitate, mainly, the dc-offset cancellation. In the TX, two-step-up is kept in use to fulfill the RX–TX turn-around time, whereas it is direct-up for b and g. No on-chip power amplifier is integrated. Two LNAs for 2.4-GHz and 5-GHz bands and two channel-selection filters (two fifth-order 25-MHz Butterworth polyphase filters centered at 10MHz are used for 802.11 a and g, other two are fifth-order 8-MHz Chebyshev LPFs centered at DC for 802.11b) are implemented with shared mixers and programmable-gain controls (PGCs). The strength of the signal is indicated by a receiver signal strength indicator (RSSI). One frequency synthesizer with VCO is used between different modes, and the receiver and transmitter. Low-dropout voltage (LDO) regulators are embedded.

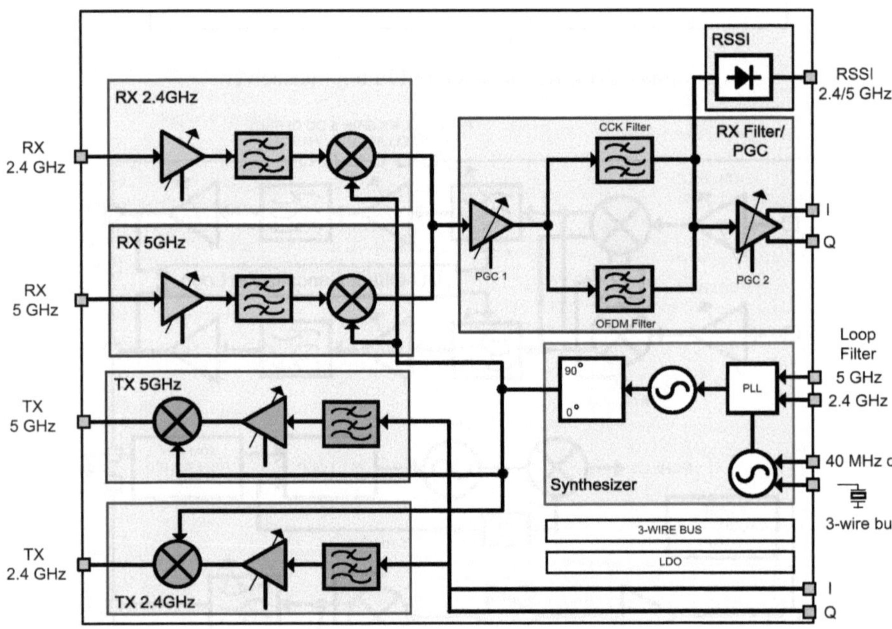

Figure 2-16. T. Rühlicke 802.11a/b/g transceiver

Discussion – Table 2-10 summarizes the performances of the three TXRs, R. Ahola's transceiver sets a common IF for 802.11a/b/g such that the problem of LO-to-LO self-mixing is reduced, whereas the IF and BB circuitries are shared. Z. Xu's transceiver directly downconverts the signal to the BB, resulting in very low power and area consumption. The problems of dc offset in RX, and LO leakage in TX, are tackled by cancellation loops. T. Rühlicke's transceiver uses a low-IF reception for 802.11a and g in OFDM mode to facilitate the dc-offset cancellation. Two channel-selection filters are required while the wideband PGCs are shared. Their results demonstrate that low power and small area can be achieved together by zero-IF RX with direct-up TX.

Table 2.10. Performance comparison of 802.11a/b/g transceivers

	R. Ahola et al.		Z. Xu et al.		T. Rühlicke et al.		
	802.11b/g	802.11a	802.11b/g	802.11a	802.11a	802.11b	802.11g
RX Architecture	Dual-Conversion		Zero-IF		Low-IF	Zero-IF	Zero-IF (CCK) Low-IF (OFDM)
TX Architecture	Dual-Conversion		Direct-Up		2-Step-Up	Direct-Up	Direct-Up
Integration	Transceiver *		Transceiver *		Transceiver *		
Technology	0.18-μm CMOS		0.18-μm CMOS		0.25-μm BiCMOS		
Supply	1.8 V		1.8 V		3 V		
Chip Area	12 mm^2		6 mm^2		13.5 mm^2		
IIP3 (min. gain)	–1 dBm	–11.8 dBm	–7.3/ 10 dBm $^{\Delta}$	–8/ 14.5 dBm$^{\Delta}$	—		
Noise Figure	5.2 dB	5.6 dB	4.9 dB	5.0 dB	—		
Sensitivity $^{+}$	—		—		–74 dBm	–85 dBm	–74 dBm
Power in RX	194.4 mW	201.6 mW	122.4 mW	115 mW	600 mW	345 mW	345 mW
Power in TX	241.2 mW	241.2 mW	136.8 mW	181.8 mW	465 mW	330 mW	330 mW

$^{\Delta}$: high/low gain, *: no A/D and D/A, $^{+}$: 54 Mb/s for a, g 11Mb/s for b

7. SUMMARY

We have summarized in this chapter some essential information that is worth considering during transceiver architecture selection. Some general conclusions are drawn below:

Architecturally, to meet the goals of low power and low cost as well as high integration and multistandard compliance, the superheterodyne architecture should be avoided. Instead, a low-IF/zero-IF-mixed RX is appeared as an effective architecture to meet those goals simultaneously, especially because the modern wireless standards consist of both narrowband and wideband applications. On the other hand, the dual-conversion architecture can cooperate with the low-IF and zero-IF ones in case of a large frequency difference in the given standards. Exploiting a common IF constitutes an efficient way to alleviate the frequency synthesis, filtering and channel selection.

In circuit level, although dedicated LNAs/PAs optimized for each band are normally required for multiband communications, inductor sharing is still efficient to save chip area. In the dual-conversion scheme, the demodulation can be performed at a common IF such that the mixers, frequency synthesizer and VCO can be shared. Since the frequency synthesizer has to provide two very different step sizes (e.g., 200 kHz and 5 MHz), in order to maintain a high reference frequency, the fractional-*N* PLL structure is more appropriate than its integer-*N* counterpart. Beside, TDMA standards allow the baseband building circuitry to be shared between TX and RX paths. A complex/real filter with tunable center-frequency and band-width can serve both low-IF/zero-IF reception and two-step-up/direct-up transmission. The bandwidth of the employed operational amplifier should be able to scale with the channel's bandwidth to save power. The A/D converter can be based on power- [2.1.15] or resolution-scalable [2.1.16] architectures.

In overall, it is obvious that realizing transceivers with high multi-standardability constitutes a very challenging task as many new wireless standards are continuously being deployed, like the recent cases of 802.11n and WiMax. They together with the existing WCDMA, 802.11a/g and UWB, are believed to be the major parts of the future wireless-ubiquitous network [2.1.17]. Developing *effective* transceiver solutions that help supporting all (most of) those standards in one terminal will be a very important direction of wireless research in the coming years.

REFERENCES

Part I

[2.1.1] T. Lee, *The Design of CMOS Radio Frequency Integrated Circuits*, Cambridge University Press, 1998.

[2.1.2] R. Hartley, "Modulation System," *U.S. Patent* 1,666,206, Apr. 1928.

[2.1.3] D. Weaver, "A Third Method of Generation and Detection of Single Sideband Signals," in *Proc. IRE*, vol. 44, no. 12, pp. 1703–1705, 1956.

[2.1.4] B. Razavi, *RF Microelectronics*, Prentice-Hall, 1998.

[2.1.5] J. Crols and M. Steyaert, *CMOS Wireless Transceiver Design*, Kluwer Academic Publishers, 1997.

[2.1.6] *Digital Cellular Telecommunications System (Phase 2); Radio Transmission and Reception*, GSM Standard 05, 1999.

[2.1.7] "UE Radio Transmission and Reception," Third Generation Partnership Project (3GPP), Doc. No. 25.101-V5.5.0.

[2.1.8] *Wireless LAN Medium Access Control (MAC) and Physical Layer (PHY) Specifications, ANSI/IEEE Standard 802.11.* [online] Available: http://standards.ieee.org/getieee802/802.11html

[2.1.9] "Broadband Radio Access Network (BRAN); HiperLAN type 2; Physical (PHY) Layer," ETSI, Sophia Antipolis Cedex, France, TS 101 475, ver.1.3.1, 2001.

[2.1.10] *Bluetooth™ v1.1 Foundation Specifications and WPAN High Rate Alternative PHY Task Group 1, IEEE 802.15.1.* [online] Available: http://www.ieee802.org/15/pub/TG1.html

[2.1.11] *HomeRF 2.01 Specification* [online] Available: http://www.palowireless.com/homerf/docs/HomeRF-2.01-us.zip

[2.1.12] *WPAN High Rate Alternative PHY Task Group 4, IEEE 802.15.4.* [online] Available: http://www.ieee802.org/15/pub/TG4.html

[2.1.13] *WPAN High Rate Alternative PHY Task Group 3a, IEEE 802.15.3a.* [online] Available: http://www.ieee802.org/15/pub/TG3a.html

[2.1.14] J. Ryynänen, K. Kivekäs, J. Jussila, L. Sumanen, A. Pärssinen and K. Halonen, "A Single-chip Multimode Receiver for GSM900, DCS1800, PCS1900, and WCDMA," *IEEE Journal of Solid-State Circuits (JSSC)*, vol. 38, no. 4, pp. 594–602, Apr. 2003.

[2.1.15] I. Ahmed and D. Johns, "A 50MS/s (35mW) to 1kS/s (15μW) Power Scaleable 10b Pipelined ADC with Minimal Bias Current Variation," *IEEE International Solid-State Circuits Conference (ISSCC), Digest of Technical Papers*, pp. 280–281, Feb. 2005.

[2.1.16] B. Xia, A. Valdes-Garcia and E. Sánchez-Sinencio, "A Configurable Time-Interleaved Pipeline ADC for Multi-Standard Wireless Receivers," in *Proc. European Solid-State Circuits Conference (ESSCIRC)*, pp. 259–262, Sept. 2004.

[2.1.17] P.-I. Mak, S.-P. U and R. P. Martins, "Transceiver Architectural Selection: Review, State-of-the-Art Survey and Case Study," *IEEE Circuits and Systems Magazine*, 2nd Issue, pp. 6–25, 2007.

Part II

GSM/DCS/PCS/GPRS – Transceiver

[2.2.1] K. Irie, et al., "A 2.7V GSM Transceiver IC," *IEEE International Solid-State Circuits Conference (ISSCC), Digest of Technical Papers*, pp. 302–303, Feb. 1997. **[SH]**

[2.2.2] M. Steyaert, et al., "A Single-Chip CMOS Transceiver for DCS-1800 Wireless Communications," *IEEE International Solid-State Circuits Conference (ISSCC), Digest of Technical Papers*, pp. 48–49, Feb. 1998. **[LIF/DU]**

[2.2.3] T. Yamawaki, et al., "A Dual-band Transceiver for GSM and DCS1800 Applications," in *Proc. European Solid-State Circuits Conference (ESSCIRC)*, pp. 84–87, Sept. 1998. **[SH/PLL]**

[2.2.4] P. Orsatti, et al., "A 20 mA-Receive 55 mA-Transmit GSM Transceiver in 0.25 μm CMOS," *IEEE International Solid-State Circuits Conference (ISSCC), Digest of Technical Papers*, pp. 232–233, Feb. 1999. **[SH]**

[2.2.5] M. Steyaert, et al., "A 2 V CMOS Cellular Transceiver Front-End," *IEEE International Solid-State Circuits Conference (ISSCC), Digest of Technical Papers*, pp. 142–143, Feb. 2000. **[LIF/DU]**

[2.2.6] S. Tanaka, et al., "GSM/DCS1800 Dual Band Direct-Conversion Transceiver IC with a DC Offset Calibration System," in *Proc. European Solid-State Circuits Conference (ESSCIRC)*, pp. 494–497, Sept. 2001. **[ZIF/DU]**

[2.2.7] S. Dow, et al., "A Dual-Band Direct-Conversion/VLIF Transceiver for 50GSM/GSM/DCS/PCS," *IEEE International Solid-State Circuits Conference (ISSCC), Digest of Technical Papers*, pp. 230–231, Feb. 2002. **[ZIF/LIF/DU]**

[2.2.8] A. Molnar, et al., "A Single-Chip Quad-Band (850/900/1800/1900MHz) Direct-Conversion GSM/GPRS RF Transceiver with Integrated VCOs and Fractional-N Synthesizer," *IEEE International Solid-State Circuits Conference (ISSCC), Digest of Technical Papers*, pp. 232–233, Feb. 2002. **[ZIF/DU]**

[2.2.9] S. Cipriani, et al., "Fully Integrated Zero IF Transceiver for GPRS/GSM/DCS/PCS Application," in *Proc. European Solid-State Circuits Conference (ESSCIRC)*, pp. 439–442, Sept. 2002. **[ZIF/DU]**

[2.2.10] E. Duvivier, et al., "A Fully Integrated Zero-IF Transceiver for GSM-GPRS Quad Band Application," *IEEE International Solid-State Circuits Conference (ISSCC), Digest of Technical Papers*, pp. 274–275, Feb. 2003. **[ZIF/DU]**

[2.2.11] E. Götz, et al., "A Quad-Band Low-Power Single-Chip Direct-Conversion CMOS Transceiver with ΣΔ-modulation loop for GSM," in *Proc. European Solid-State Circuits Conference (ESSCIRC)*, pp. 217–220, Sept. 2003. **[ZIF/DU]**

[2.2.12] J. Rudell, et al., "A Single-Chip Quad-Band GSM/GPRS Transceiver in 0.18μm Standard CMOS," *IEEE International Solid-State Circuits Conference (ISSCC), Digest of Technical Papers*, pp. 318–319, Feb. 2005. **[ZIF/PLL]**

[2.2.13] R. Magoon, et al., "A 900MHz/1.9GHz Integrated Transceiver and Synthesizer IC for GSM," in *Proc. European Solid-State Circuits Conference (ESSCIRC)*, pp. 53–56, Sept. 2000. **[SH]**

GSM/DCS/PCS/GPRS – Receiver

[2.2.14] S. Tadjpour, E. Cijvat, E. Hegazi and A. Abidi, "A 900MHz Dual Conversion Low-IF GSM Receiver in 0.35 μm CMOS," *IEEE International Solid-State Circuits Conference (ISSCC), Digest of Technical Papers*, pp. 292–293, Feb. 2001. **[LIF]**

[2.2.15] B. Razavi, et al., "A 900-MHz CMOS Direct Conversion Receiver," *IEEE Symposium on VLSI Circuits (VLSI), Digest of Technical Papers*, pp. 113–114, June 1997. **[ZIF]**

GSM/DCS/PCS/GPRS – Transmitter

[2.2.16] B. Razavi, et al., "A 900-MHz/1.8-GHz CMOS Transmitter for Dual-band Applications," *IEEE Symposium on VLSI Circuits (VLSI), Digest of Technical Papers*, pp. 128–131, June 1998. **[2SU]**

[2.2.17] M. Borremans and M. Steyaert, "A CMOS 2V Quadrature Direct Up-Converter Chip for DCS-1800 Integration," in *Proc. European Solid-State Circuits Conference (ESSCIRC)*, pp. 49–52, Sept. 2000. **[DU]**

[2.2.18] E. Hegazi and A. A. Abidi, "A 17mW Transmitter and Frequency Synthesizer for 900MHz GSM Fully Integrated in 0.35-μm CMOS," *IEEE Symposium on VLSI Circuits (VLSI), Digest of Technical Papers*, pp. 234–237, June 2002. **[PLL]**

[2.2.19] T. Sowlati, et al., "Quad-Band GSM/GPRS/EDGE Polar Loop Transmitter," *IEEE International Solid-State Circuits Conference (ISSCC), Digest of Technical Papers*, pp. 186–187, Feb. 2004. **[POLAR]**

[2.2.20] S. T. Lee, et al., "A 1.5V 28mA Fully-Integrated Fast-Locking Quad-Band GSM-GPRS Transmitter with Digital Auto-Calibration in 130nm CMOS," *IEEE International Solid-State Circuits Conference (ISSCC), Digest of Technical Papers*, pp. 188–189, Feb. 2004. **[PLL]**

[2.2.21] M. Elliott, et al., "A Polar Modulator Transmitter for EDGE," *IEEE International Solid-State Circuits Conference (ISSCC), Digest of Technical Papers*, pp. 190–191, Feb. 2004. **[POLAR]**

[2.2.22] A. Hadjichristos, et al., "A highly Integrated Quad Band Low EVM Polar Modulation Transmitter for GSM/EDGE Applications," in *Proc. IEEE Custom Integrated-Circuit Conference (CICC)*, pp. 565–568, Oct. 2004. **[POLAR]**

WCDMA – Transceiver

[2.2.23] T. Maruyama, et al., "Single-Chip IF Transceiver IC with Wide Dynamic Range Variable Gain Amplifiers for Wideband CDMA Applications," *IEEE Symposium on VLSI Circuits (VLSI), Digest of Technical Papers*, pp. 11–14, June 2001. **[SH]**

[2.2.24] B. Pellat, et al., "Fully-Integrated WCDMA SiGeC BiCMOS Transceiver," in *Proc. European Solid-State Circuits Conference (ESSCIRC)*, pp. 519–522, Sept. 2005. **[ZIF/DUAL]**

WCDMA – Receiver

[2.2.25] A. Pärssinen, et al., "A Wide-Band Direct Conversion Receiver for WCDMA Applications," *IEEE International Solid-State Circuits Conference (ISSCC), Digest of Technical Papers*, pp. 220–221, Feb. 1999. **[ZIF]**

[2.2.26] A. Pärssinen, et al., "A Wide-Band Direct Conversion Receiver with On-Chip A/D Converters," *IEEE Symposium on VLSI Circuits (VLSI), Digest of Technical Papers*, pp. 32–33, June 2000. **[ZIF]**

[2.2.27] D. Yee, et al., "A 2-GHz Low-Power Single-Chip CMOS Receiver for WCDMA Applications," in *Proc. European Solid-State Circuits Conference (ESSCIRC)*, pp. 57–60, Sept. 2000. **[ZIF]**

[2.2.28] J. Jussila, et al., "A 22mA 3.7dB NF Direct Conversion Receiver for 3G WCDMA," *IEEE International Solid-State Circuits Conference (ISSCC), Digest of Technical Papers*, pp. 284–285, Feb. 2001. **[ZIF]**

[2.2.29] M. Ugajin, J. Kodate and T. Tsukahara, "A 1V 12mW 2GHz Receiver with 49dB Image Rejection in CMOS/SIMOX," *IEEE International Solid-State Circuits Conference (ISSCC), Digest of Technical Papers*, pp. 288–289, Feb. 2001. **[LIF]**

[2.2.30] L. Der and B. Razavi, "A 2GHz CMOS Image-Reject Receiver with Sign-Sign LMS Calibration," *IEEE International Solid-State Circuits Conference (ISSCC), Digest of Technical Papers*, pp. 294–295, Feb. 2001. **[WEAVER]**

[2.2.31] R. Gharpurey, et al., "A Direct Conversion Receiver for the 3G WCDMA Standard," in *Proc. IEEE Custom Integrated-Circuit Conference (CICC)*, pp. 239–242, May 2002. **[ZIF]**

[2.2.32] J. Rogin, I. Kouchev and Q. Huang, "A 1.5V 45mW Direct Conversion WCDMA Receiver IC in 0.13μm CMOS," *IEEE International Solid-State Circuits Conference (ISSCC), Digest of Technical Papers*, pp. 268–269, Feb. 2003. **[ZIF]**

[2.2.33] E. Sacchi, et al., "A 15 mW, 70 kHz 1/f Corner Direct Conversion CMOS Receiver," in *Proc. IEEE Custom Integrated-Circuit Conference (CICC)*, pp. 459–462, Sept. 2003. [WCDMA] **[ZIF]**

[2.2.34] J. Ryynanen, et al., "WCDMA Multicarrier Receiver for Base-Station Applications," in *Proc. European Solid-State Circuits Conference (ESSCIRC)*, pp. 515–518, Sept. 2005. **[LIF]**

WCDMA – Transmitter

[2.2.35] A. Bellaouar, et al., "A Highly-Integrated SiGe BiCMOS WCDMA Transmitter IC," *IEEE International Solid-State Circuits Conference (ISSCC), Digest of Technical Papers*, pp. 238–239, Feb. 2002. **[SH]**

[2.2.36] G. Brenna, D. Tschopp and Q. Huang, "Carrier Leakage Suppression in Direct-Conversion WCDMA Transmitters," *IEEE International Solid-State Circuits Conference (ISSCC), Digest of Technical Papers*, pp. 270–271, Feb. 2003. **[DU]**

[2.2.37] V. Leung, et al., "Digital-IF WCDMA Handset Transmitter IC in 0.25μm SiGe BiCMOS," *IEEE International Solid-State Circuits Conference (ISSCC), Digest of Technical Papers*, pp. 182–183, Feb. 2004. **[2SU]**

802.11b – SoC

[2.2.38] H. Darabi, et al., "A Fully Integrated SoC for 802.11b in 0.18μm CMOS," *IEEE International Solid-State Circuits Conference (ISSCC), Digest of Technical Papers*, pp. 96–97, Feb. 2005. **[ZIF/DU]**

802.11b – Transceiver

[2.2.39] P. Stroet, et al., "A Zero-IF Single-Chip Transceiver for up to 22 Mb/s QPSK 802.11b Wireless LAN," *IEEE International Solid-State Circuits Conference (ISSCC), Digest of Technical Papers*, pp. 204–205, Feb. 2001. **[ZIF/DU]**

[2.2.40] G. Chien, et al., "A 2.4GHz CMOS Transceiver and Baseband Processor Chipset for 802.11b Wireless LAN Application," *IEEE International Solid-State Circuits Conference (ISSCC), Digest of Technical Papers*, pp. 358–359, Feb. 2003. **[SH]**

[2.2.41] W. Kluge, et al., "A 2.4GHz CMOS Transceiver for 802.11b Wireless LANs," *IEEE International Solid-State Circuits Conference (ISSCC), Digest of Technical Papers*, pp. 360–361, Feb. 2003. **[ZIF/DU]**

802.11b – Receiver

[2.2.42] F. Behbahani, et al., "An Adaptive 2.4 GHz Low-IF Receiver in 0.6-μm CMOS for Wideband Wireless LAN," *IEEE International Solid-State Circuits Conference (ISSCC), Digest of Technical Papers*, pp. 146–147, Feb. 2000. **[LIF]**

802.11a (HiperLAN 2) – Transceiver

[2.2.43] J. Plouchart, H. Ainspan and M. Soyuer, "A 5.2 GHz 3.3V I/Q SiGe RF Transceiver," in *Proc. IEEE Custom Integrated-Circuit Conference (CICC)*, pp. 217–220, May 1999. **[HIF]**

[2.2.44] T.-P. Liu, et al., "5 GHz CMOS Radio Transceiver Front-End Chipset," *IEEE International Solid-State Circuits Conference (ISSCC), Digest of Technical Papers*, pp. 320–321, Feb. 2000. **[ZIF/DU]**

[2.2.45] D. Su, et al., "A 5GHz CMOS Transceiver for IEEE 802.11a Wireless LAN," *IEEE International Solid-State Circuits Conference (ISSCC), Digest of Technical Papers*, pp. 92–93, Feb. 2002. **[DUAL]**

[2.2.46] T. Schwanenberger, M. Ipek, S. Roth and H. Schemmann, "A Multistandard Single-Chip Transceiver Covering 5.15 to 5.85GHz," *IEEE International Solid-State Circuits Conference (ISSCC), Digest of Technical Papers*, pp. 350–351, Feb. 2003. **[SH]**

[2.2.47] I. Bouras, et al., "A Digitally Calibrated 5.15 - 5.825GHz Transceiver for 802.11a Wireless LANs in 0.18μm CMOS," *IEEE International Solid-State Circuits Conference (ISSCC), Digest of Technical Papers*, pp. 352–353, Feb. 2003. **[ZIF/DU]**

[2.2.48] P. Zhang, et al., "A Direct Conversion CMOS Transceiver for IEEE 802.11a WLANs," *IEEE International Solid-State Circuits Conference (ISSCC), Digest of Technical Papers*, pp. 354–355, Feb. 2003. **[ZIF/DU]**

[2.2.49] A. Behzad, et al., "Direct-Conversion CMOS Transceiver with Automatic Frequency Control for 802.11a Wireless LANs," *IEEE International Solid-State Circuits Conference (ISSCC), Digest of Technical Papers*, pp. 356–357, Feb. 2003. **[ZIF/DU]**

[2.2.50] T. Maeda, et al., "A Direct-Conversion CMOS Transceiver for 4.9–5.95 GHz Multistandard WLANs," *IEEE International Solid-State Circuits Conference (ISSCC), Digest of Technical Papers*, pp. 90–91, Feb. 2004. **[ZIF/DU]**

802.11a (HiperLAN 2) – Receiver

[2.2.51] J. Maligeorgos and J. Long, "A 2 V 5.1–5.8 GHz Image-Reject Receiver with Wide Dynamic Range," *IEEE International Solid-State Circuits Conference (ISSCC), Digest of Technical Papers*, pp. 322–323, Feb. 2000. **[HIF]**

[2.2.52] B. Razavi, et al., "A 5.2-GHz CMOS Receiver with 62-dB Image Rejection," *IEEE Symposium on VLSI Circuits (VLSI), Digest of Technical Papers*, pp. 34–37, June 2000. **[DUAL]**

[2.2.53] H. Samavati, H. Rategh and T. Lee, "A Fully Integrated 5GHz CMOS Wireless LAN Receiver," *IEEE International Solid-State Circuits Conference (ISSCC), Digest of Technical Papers*, pp. 208–209, Feb. 2001. **[DUAL]**

[2.2.54] C.-Y. Wu and C.-Y. Chou, "A 5-GHz CMOS Double-Quadrature Receiver for IEEE 802.11a Applications," *IEEE Symposium on VLSI Circuits (VLSI), Digest of Technical Papers*, pp. 149–152, June 2003. **[LIF]**

[2.2.55] G. Montagna, et al., "A 72mW CMOS 802.11a Direct Conversion Receiver with 3.5dB NF and 200kHz 1/f Noise Corner," *IEEE Symposium on VLSI Circuits (VLSI), Digest of Technical Papers*, pp. 16–19, June 2004. **[ZIF]**

802.11g – SoC

[2.2.56] S. Mehta, et al., "An 802.11g WLAN SoC," *IEEE International Solid-State Circuits Conference (ISSCC), Digest of Technical Papers*, pp. 94–95, Feb. 2005. **[ZIF/DU]**

802.11g – Transceiver

[2.2.57] Y. Hsieh, et al., "An Auto-I/Q Calibrated CMOS Transceiver for 802.11g," *IEEE International Solid-State Circuits Conference (ISSCC), Digest of Technical Papers*, pp. 92–93, Feb. 2005. **[ZIF/DU]**

Bluetooth – SoC

[2.2.58] F. Eynde, et al., "A Fully-Integrated Single-Chip SOC for Bluetooth," *IEEE International Solid-State Circuits Conference (ISSCC), Digest of Technical Papers*, pp. 196–197, Feb. 2001. **[LIF/DU]**

[2.2.59] J. Cheah, et al., "Design of a Low-Cost Integrated 0.25 μm CMOS Bluetooth SOC in 16.5 mm^2 Silicon Area," *IEEE International Solid-State Circuits Conference (ISSCC), Digest of Technical Papers*, pp. 90–91, Feb. 2002. **[LIF/PLL]**

Bluetooth – Transceiver

[2.2.60] A. Ajjikuttira, et al., "A Fully-Integrated CMOS RFIC for Bluetooth Applications," *IEEE International Solid-State Circuits Conference (ISSCC), Digest of Technical Papers*, pp. 198–199, Feb. 2001. **[LIF/DU]**

[2.2.61] H. Darabi, et al., "A 2.4 GHz CMOS Transceiver for Bluetooth," *IEEE International Solid-State Circuits Conference (ISSCC), Digest of Technical Papers*, pp. 200–201, Feb. 2001. **[LIF/DU]**

[2.2.62] H. Komurasaki, et al., "A Single-Chip 2.4 GHz RF Transceiver LSI with a Wide-Range FV Conversion Demodulator," *IEEE International Solid-State Circuits Conference (ISSCC), Digest of Technical Papers*, pp. 206–207, Feb. 2001. **[LIF/PLL]**

[2.2.63] N. Filiol, et al., "A 22 mW Bluetooth RF Transceiver with Direct RF Modulation and On-Chip IF Filtering," *IEEE International Solid-State Circuits Conference (ISSCC), Digest of Technical Papers*, pp. 202–203, Feb. 2001. **[LIF/PLL]**

[2.2.64] S.-W. Lee, et al., "A Single-Chip 2.4GHz Direct-Conversion CMOS Transceiver with GFSK Modem for Bluetooth Application," *IEEE Symposium on VLSI Circuits (VLSI), Digest of Technical Papers*, pp. 245–246, June 2001. **[ZIF/DU]**

[2.2.65] P. van Zeijl, et al., "A Bluetooth Radio in 0.18 µm CMOS," *IEEE International Solid-State Circuits Conference (ISSCC), Digest of Technical Papers*, pp. 86–87, Feb. 2002. **[LIF/DU]**

[2.2.66] G. Chang, et al., "A Direct-Conversion Single-Chip Radio-Modem for Bluetooth," *IEEE International Solid-State Circuits Conference (ISSCC), Digest of Technical Papers*, pp. 88–89, Feb. 2002. **[ZIF/DU]**

[2.2.67] M. Kokuko, et al., "A 2.4 GHz RF Transceiver with Digital Channel-Selection Filter for Bluetooth," *IEEE International Solid-State Circuits Conference (ISSCC), Digest of Technical Papers*, pp. 93–94, Feb. 2002. **[LIF/PLL]**

[2.2.68] H. Komurasaki, et al., "A 1.8-V Operation RFCMOS Transceiver for Bluetooth," *IEEE Symposium on VLSI Circuits (VLSI), Digest of Technical Papers*, pp. 230–233, June 2002. **[LIF/PLL]**

[2.2.69] C. Cojocaru, et al., "A 43mW Bluetooth Transceiver with –91dBm Sensitivity," *IEEE International Solid-State Circuits Conference (ISSCC), Digest of Technical Papers*, pp. 90–91, Feb. 2003. **[LIF/PLL]**

[2.2.70] H. Ishikuro, et al., "A Single-Chip CMOS Bluetooth Transceiver with 1.5MHz IF and Direct Modulation Transmitter," *IEEE International Solid-State Circuits Conference (ISSCC), Digest of Technical Papers*, pp. 94–95, Feb. 2003. **[LIF/PLL]**

[2.2.71] C.-H. Park, et al., "A Low Power CMOS Bluetooth Transceiver with a Digital Offset Canceling DLL-Based GFSK Demodulator," *IEEE International Solid-State Circuits Conference (ISSCC), Digest of Technical Papers*, pp. 96–97, Feb. 2003. **[LIF/DU]**

[2.2.72] M. Ugajin, et al., "A 1-V CMOS/SOI Bluetooth RF Transceiver for Compact Mobile Applications," *IEEE Symposium on VLSI Circuits (VLSI), Digest of Technical Papers*, pp. 123–126, June 2003. **[SH/PLL]**

[2.2.73] M. Chen, et al., "A CMOS Bluetooth Radio Transceiver using a Sliding-IF Architecture," in *Proc. IEEE Custom Integrated-Circuit Conference (CICC)*, pp. 455–458, Sept. 2003. **[DUAL]**

Bluetooth – Receiver

[2.2.74] B. Razavi, et al., "A 900-MHz CMOS Direct Conversion Receiver," *IEEE Symposium on VLSI Circuits (VLSI), Digest of Technical Papers*, pp. 113–114, June 1997. **[ZIF]**

[2.2.75] W. Sheng, et al., "A Monolithic CMOS Low-IF Bluetooth Receiver," in *Proc. IEEE Custom Integrated-Circuit Conference (CICC)*, pp. 247–250, May 2002. **[LIF]**

[2.2.76] K. Muhammad, et al., "A Discrete-Time Bluetooth Receiver in a 0.13µm Digital CMOS Process," *IEEE International Solid-State Circuits Conference (ISSCC), Digest of Technical Papers*, pp. 268–269, Feb. 2004. **[LIF]**

Bluetooth – Transmitter

[2.2.77] D. Miyashita, et al., "A Low-IF CMOS Single-Chip Bluetooth EDR Transmitter with Digital I/Q Mismatch Trimming Circuit," *IEEE Symposium on VLSI Circuits (VLSI), Digest of Technical Papers*, pp. 299–301, June 2005. **[2SU]**

802.15.4 – Transceiver

[2.2.78] P. Choi, et al., "An Experimental Coin-Sized Radio for Extremely Low Power WPAN (IEEE802.15.4) Application at 2.4GHz," *IEEE International Solid-State Circuits Conference (ISSCC), Digest of Technical Papers*, pp. 92–93, Feb. 2003. **[LIF/DU]**

UWB – Transceiver

[2.2.79] S. Lida, et al., "A 3.1 to 5GHz CMOS DSSS UWB Transceiver for WPANs," *IEEE International Solid-State Circuits Conference (ISSCC), Digest of Technical Papers*, pp. 214–215, Feb. 2005. **[ZIF/DU]**

[2.2.80] B. Razavi, et al., "A 0.13μm CMOS UWB Transceiver," *IEEE International Solid-State Circuits Conference (ISSCC), Digest of Technical Papers*, pp. 216–217, Feb. 2005. **[ZIF/DU]**

UWB – Receiver

[2.2.81] A. Ismail and A. Abidi, "An Interference Robust Receive Chain for UWB Radio in SiGe BiCMOS," *IEEE International Solid-State Circuits Conference (ISSCC), Digest of Technical Papers*, pp. 200–201, Feb. 2005. **[ZIF]**

[2.2.82] A. Ismail and A. Abidi, "A 3.1 to 8.2 GHz Direct Conversion Receiver for MB-OFDM UWB Communications," *IEEE International Solid-State Circuits Conference (ISSCC), Digest of Technical Papers*, pp. 208–209, Feb. 2005. **[ZIF]**

UWB – Transmitter

[2.2.83] S. Aggarwal, et al., "A Low Power Implementation for the Transmit Path of a UWB Transceiver," in *Proc. IEEE Custom Integrated-Circuit Conference (CICC)*, pp. 149–152, Sept. 2005. **[DU]**

CDMA/WCDMA/AMPS/GPS – Receiver

[2.2.84] V. Aparin, et al., "A Highly-Integrated Tri-Band/Quad-Mode SiGe BiCMOS RF-to-Baseband Receiver for Wireless CDMA/WCDMA/AMPS Applications with GPS Capability," *IEEE International Solid-State Circuits Conference (ISSCC), Digest of Technical Papers*, pp. 234–235, Feb. 2002. **[SH]**

CDMA/WCDMA/AMPS/GPS – Transmitter

[2.2.85] K. Gard, et al., "Direct Conversion Dual-Band SiGe BiCMOS Transmitter and Receive PLL IC for CDMA/WCDMA/AMPS/GPS Applications," *IEEE international Solid-State Circuits Conference (ISSCC), Digest of Technical Papers*, pp. 272–273, Feb. 2003. **[DU]**

Bluetooth and 802.11b – Transceiver

[2.2.86] H. Darabi, et al., "A Dual Mode 802.11b/Bluetooth Radio in 0.35μm CMOS," *IEEE International Solid-State Circuits Conference (ISSCC), Digest of Technical Papers*, pp. 86–87, Feb. 2003. **[ZIF/LIF/DU]**

[2.2.87] T. Cho, et al., "A 2.4GHz Dual-Mode 0.18μm CMOS Transceiver for Bluetooth and 802.11b," *IEEE International Solid-State Circuits Conference (ISSCC), Digest of Technical Papers*, pp. 88–89, Feb. 2003. **[ZIF/LIF/DU]**

[2.2.88] Y.-J. Jung, et al., "A Dual-Mode Direct-Conversion CMOS Transceiver for Bluetooth and 802.11b," in *Proc. European Solid-State Circuits Conference (ESSCIRC)*, pp. 225–228, Sept. 2003. **[ZIF/DU]**

Bluetooth and 802.11b – Receiver

[2.2.89] A. Emira, et al., "A Dual-Mode 802.11b/Bluetooth Receiver in 0.25μm BiCMOS," *IEEE International Solid-State Circuits Conference (ISSCC), Digest of Technical Papers*, pp. 270–271, Feb. 2004. **[ZIF]**

802.11a/b/g – Transceiver

[2.2.90] R. Ahola, et al., "A Single-Chip CMOS Transceiver for 802.11a/b/g Wireless LANs," *IEEE International Solid-State Circuits Conference (ISSCC), Digest of Technical Papers*, pp. 92–93, Feb. 2004. **[ZIF/DU]**

[2.2.91] L. Perraud, et al., "A Dual-Band 802.11a/b/g Radio in 0.18-μm CMOS," *IEEE International Solid-State Circuits Conference (ISSCC), Digest of Technical Papers*, pp. 94–95, Feb. 2004. **[DUAL]**

[2.2.92] M. Zargari, et al., "A Single-Chip Dual-Band Tri-Mode CMOS Transceiver for IEEE 802.11a/b/g WLAN, *IEEE International Solid-State Circuits Conference (ISSCC), Digest of Technical Papers*, pp. 96–97, Feb. 2004. **[DUAL]**

[2.2.93] Z. Xu, et al., "A Compact Dual-Band Direct-Conversion CMOS Transceiver for 802.11a/b/g WLAN, *IEEE International Solid-State Circuits Conference (ISSCC), Digest of Technical Papers*, pp. 98–99, Feb. 2005. **[ZIF/DU]**

[2.2.94] T. Maeda, et al., "A Low-Power Dual-Band Triple-Mode WLAN CMOS Transceiver," *IEEE International Solid-State Circuits Conference (ISSCC), Digest of Technical Papers*, pp. 100–101, Feb. 2005. **[ZIF/DU]**

[2.2.95] A. Ravi, et al., "A 1.4V, 2.4/5 GHz, 90nm CMOS System in a Package Transceiver for Next Generation WLAN," *IEEE Symposium on VLSI Circuits (VLSI), Digest of Technical Papers*, pp. 294–297, June 2005. **[ZIF/DU]**

[2.2.96] N. Haralabidis, et al., "A Cost-Efficient 0.18μm CMOS RF Transceiver using a Fractional-N Synthesizer for 802.11b/g Wireless LAN Applications," in *Proc. IEEE Custom Integrated-Circuit Conference (CICC)*, pp. 405–408, Oct. 2004. **[ZIF/DU]**

[2.2.97] P. Zhang, et al., "A CMOS Direct-Conversion Transceiver for IEEE 802.11a/b/g WLANs," in *Proc. IEEE Custom Integrated-Circuit Conference (CICC)*, pp. 409–412, Oct. 2004. **[ZIF/DU]**

[2.2.98] K. Vavelidis, et al., "A Single-Chip, 5.15GHz-5.35GHz, 2.4GHz-2.5GHz, 0.18μm CMOS RF Transceiver for 802.11a/b/g Wireless LAN," in *Proc. European Solid-State Circuits Conference (ESSCIRC)*, pp. 221–224, Sept. 2003. **[ZIF/DU]**

[2.2.99] T. Rühlicke, M. Zannoth and B. Klepser, "A Highly Integrated, Dual-Band, Multi-Mode Wireless LAN Transceiver," in *Proc. European Solid-State Circuits Conference (ESSCIRC)*, pp. 229–232, Sept. 2003. **[ZIF/LIF/2SU/DU]**

[2.2.100]J. Rogers, et al., "A Fully Integrated Multi-Band MIMO WLAN Transceiver RFIC," *IEEE Symposium on VLSI Circuits (VLSI), Digest of Technical Papers*, pp. 290–293, June 2005. **[DUAL]**

[2.2.101]O. Charlon, et al., "A Low-Power High-Performance SiGe BiCMOS 802.11a/b/g Transceiver IC for Cellular and Bluetooth Co-Existence Applications," in *Proc. European Solid-State Circuits Conference (ESSCIRC)*, pp. 129–132, Sept. 2005. **[ZIF/DU]**

Chapter 3

TWO-STEP CHANNEL SELECTION – A TECHNIQUE FOR MULTISTANDARD TRANSCEIVER FRONT-ENDS

1. INTRODUCTION

This chapter starts with a review of the prevailing channel-selection techniques utilized so far in the design of wireless transceiver analog front-ends (AFEs), followed by the presentation of a novel *Two-Step Channel-Selection* technique. It handles the traditionally unwanted image, in RF-to-IF (or IF-to-RF) frequency conversion, as a useful adjacent of the desired channel, and selects deliberately either of them from IF to baseband (or baseband to IF). Thus, one additional possibility for selecting channels will be created for both low-IF receivers and two-step-up transmitters, resulting in two types of benefits. First, many design specifications of the RF frequency synthesizer (e.g., settling time) and local oscillator (e.g., phase noise) can be substantially relaxed. Second, a low-IF/zero-IF reconfigurable receiver and a direct-up/two-step-up reconfigurable transmitter can be synthesized to match better with narrowband-wideband-mixed multistandard systems. The operating principles of such architectures are presented in easy-to-understand complex-signal spectral-flow (CSSF) illustrations, and their practicability is demonstrated in the design of a Bluetooth/IEEE 802.11FH/ HomeRF multistandard receiver. SPECTRE simulation results validate the reconfigurable functionalities. The implementations are based on a triple-mode channel-selection filter and a multifunctional sampling-mixer scheme, which are also presented.

2. CONVENTIONAL AND PROPOSED CHANNEL-SELECTION SCHEMES

Almost all voice- and data-centric standards utilize (or partially utilize) frequency-division multiple-access (FDMA) to divide the entire frequency bands into channels for multiple users. The mission of the AFE is to retrieve the sought channel from the air, amplifying it and downconverting it from RF to baseband for demodulation. This process is well known and widely implemented in superheterodyne receivers, namely, the sought channel is gradually downconverted and filtered from RF to different IFs, and finally to the baseband. On the other hand, image-rejection receivers use a series of steps for channel selection, which usually comprise the combination (with possible permutations) of the three main blocks, the frequency synthesizer (FS), the local oscillator (LO) and the channel-selection filter (CSF). Depending on the operating frequency (i.e., RF or IF) and movability of the blocks, it will typically lead to the subsequent two alternative architectures.

2.1 Conventional: fixed LO_{RF} + varying IF

The first type of architecture [3.1] is depicted in Figure 3-1(a), where a fixed-frequency RF I/Q local oscillator (I/Q-LO_{RF}) is exploited to perform a large step of RF-to-IF downconversion. After that, the desired channel is extracted at a relatively low-IF value by using a center-frequency-controllable CSF. Ultimately, the sought channel can be downconverted to the baseband by utilizing an IF FS and IF I/Q-LO. This structure, first, highly relaxes the phase-noise requirement of the I/Q-LO_{RF} because it is free from locking. Second, since the channel-selection filtering is performed prior to the IF-to-baseband downconversion, the operating frequency and the phase-noise requirement of the IF FS and IF I/Q-LO can be highly reduced. However, the main bottleneck of this permutation is the required broadband-tunable filter, which requires accurate control of the center frequency. For instance, in Bluetooth, if the entire band (totally 79 channels) is down-converted to baseband in the first mixing, a 1-MHz bandpass filter with 79 different center frequencies in a range of 80 MHz (–40 to +40 MHz) is needed. Moreover, the agility of the filter should be high to perform also frequency hopping. With such rigid constraints, it would be infeasible to apply this method in modern applications. However, a special case of this architecture was reported for DECT application, namely wideband IF double-conversion receiver [3.2], which employs a fixed-frequency LO incorporated with a wideband lowpass filter (LPF) in the first down-conversion, whereas the channel selection is shifted to the IF. In this way, the operating frequency of the succeeding stages is reduced, but this benefit comes at the expense of an increased linearity requirement from the wide-band LPF to prevent channel-to-channel intermodulation.

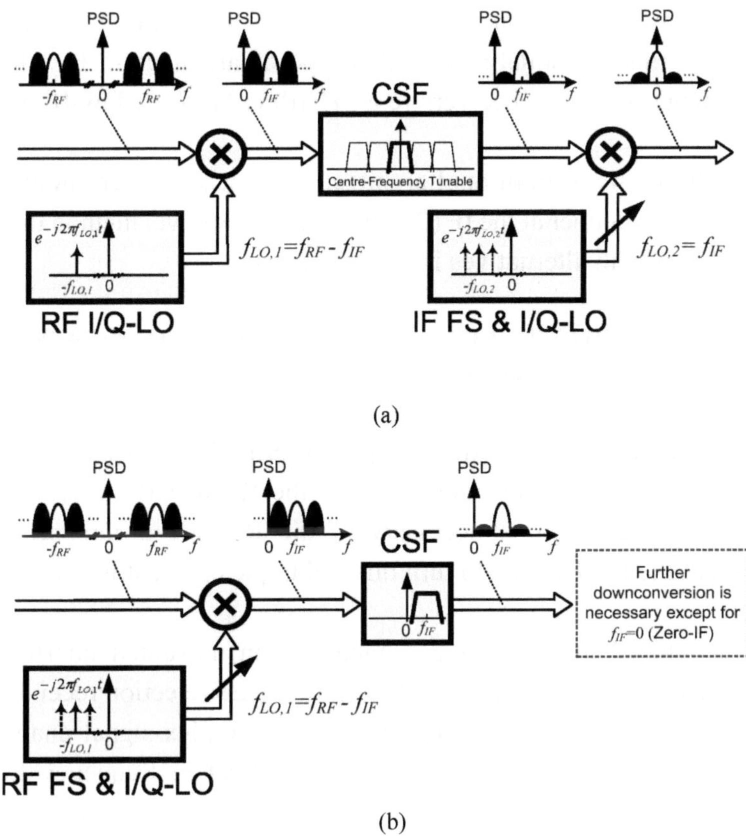

Figure 3-1. CSSF illustrations of conventional channel-selection methods: (a) fixed LO_{RF} + varying IF (b) varying LO_{RF} + fixed IF

2.2 Conventional: varying LO_{RF} + fixed IF

A second alternative architecture [3.3] is shown in Figure 3-1(b) that uses a RF FS and a I/Q-LO_{RF} to cover all the possible channel positions in the RF band of interest, then the desired channel is downconverted to the baseband, at which only a fixed CSF would be needed. This structure is well appropriate for state-of-the-art IC designs since the current developments of FSs (based on PLL architectures) have presented results of operating frequencies in the multiple-GHz range with adequate performance. On the other hand, a fast-settling and broadband-tunable oscillator is much easier to implement than its filter counterpart, and a baseband filter is much simpler and more power-efficient than a bandpass. The exhibition of these compromised features confirms the suitability of this type of architecture for almost all kinds of image-rejection receivers (e.g., Hartley, Weaver, low-IF and zero-IF) [3.3].

Concisely, the two traditional architectures just presented include movable circuit blocks either at the IF (*2.1*) or RF (*2.2*). A novel method that efficiently combines both alternatives is described next.

2.3 Proposed two-step channel selection: coarse-varying LO_{RF} + fine-varying IF

The two-step channel-selection [3.4][3.5][3.6][3.7], as its name implies, is to partition the channel selection between the RF and IF AFEs, such that only a *coarse* selection is necessary at the RF and a *fine* selection is completed at the IF. The key to turn this technique efficient is a new concept of image.

It is recognized that the image channel is an unwanted interference in frequency conversion (either up or down). In image-rejection receivers, after a quadrature downconversion and based on a complex-signal analysis, the desired channel and its image (twice the IF offset in frequency from the desired channel) will be located at the same IF but with a complex conjugate representation. Now, suppose the IF AFE can flexibly select either of them, a channel selection is accomplished without any prerequisite needed in the radio part. To implement such a technique, some specific IF values, i.e., n+0.5 channel spacing (CS) for n=0, 1, 2, 3..., should be used to select one of the adjacent channels as the image of the desired. For instance, Figure 3-2 can describe the entire operation if half (i.e., 0.5) channel spacing is selected. Such operation has two different IF-to-baseband operation modes, labeled as A and B, where the channel-selection filter has a tunable center frequency at either +IF or –IF, while the IF I/Q-LO needs to provide not only the conventional 0° and 90°, but also 180° and 270°, to make a selection between the upper and lower sidebands. The A/D can be placed prior or after the secondary mixing to trade the A/D conversion rate with the mixer implementation (i.e., analog or digital). The final shared operations are decimation and image elimination through a simple digital filter.

As introduced next, the technique required IF values that will not pose any limitation, rather, it has other advantageous features. For the reconfigurable functionalities, only simple analog circuitry with digital control is needed.

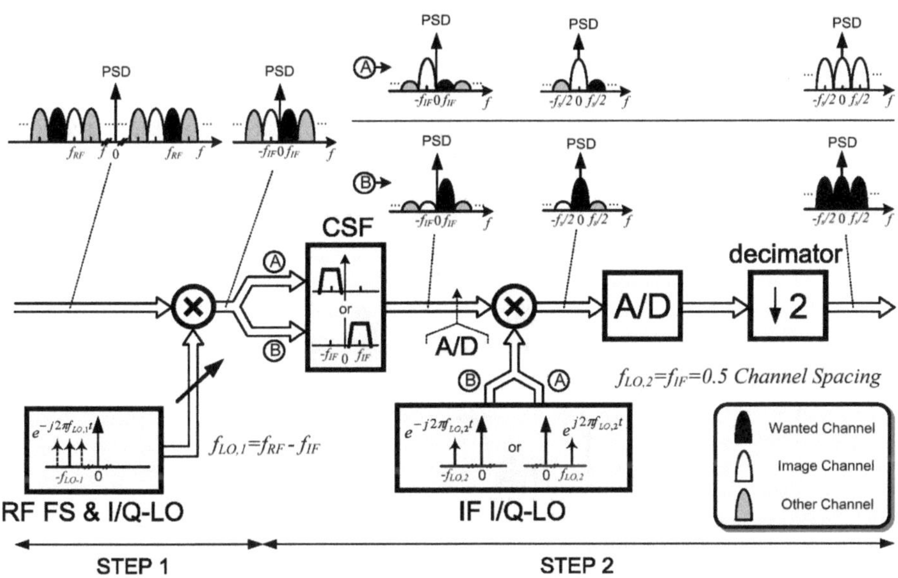

Figure 3-2. CSSF illustration of low-IF receiver with 2-SCS technique, where the A/D can be placed before or after the second mixing

IF Selection – when IF=0 (i.e., zero IF), the image channel is the sideband of the desired signal itself as shown in Figure 3-3(a), although the image-rejection requirement is much relaxed, the low-frequency disturbance becomes very serious. Conversely, when the IF value gets higher, the required image rejection needs to be increased by the power difference between the image and the desired channel. To establish a good compromise between the low-frequency and image interference, an IF equals to half of the CS can be chosen, as shown in Figure 3-3(b). This situation is especially true since in most wireless communication standards, the power of the adjacent channels increases with their frequency offset. For example, in GSM (Bluetooth), the power of the first adjacent channel is only 18 dB (5 dB) higher than the desired channel, the required image rejection is therefore only ~32 dB (~20 dB). Furthermore, the local oscillator will be locked in between every channel, and any unwanted LO leakage (e.g., through the substrate) will not degrade the signal quality. For the succeeding filter and A/D, however, their operating frequency has to be increased with the IF, implying higher power consumption. In general, only up to four channel spacings will be used as the IF in image-rejection receivers. This indicates that the proposed "n+0.5 channel spacing IF" with n = 0, 1, 2 or 3 would be feasible, and will not complicate the design.

An interesting similarity of this technique to the hierarchical-QAM scheme [3.8] is that the sideband selection features in complex-to-complex frequency down/upconversion is employed at baseband, even though the objectives, implementation and resulting advantages are differed from each other.

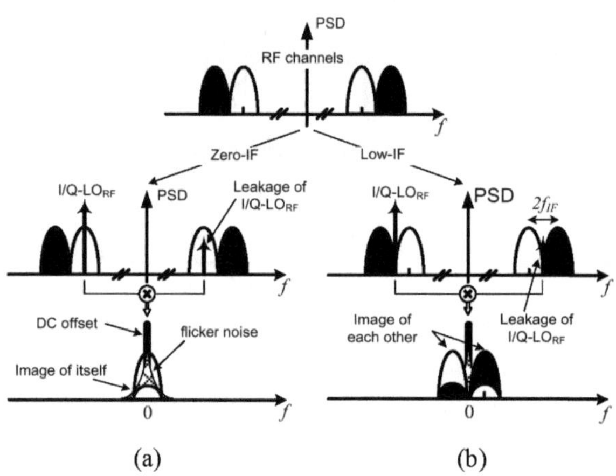

Figure 3-3. CSSF illustration of interference locations (with also I/Q mismatch): (a) IF=0 (b) IF=0.5 channel spacing

Functional blocks for second step – as mentioned before a RF FS can effectively be used for the first step. For the second step, we propose to implement the CSF by slightly modifying the structure of a conventional complex (or polyphase) filter, while the four-phased IF I/Q-LO can be replaced by a sampling-mixer scheme embedding a programmable analog double-quadrature sampling (A-DQS) technique [3.4]. With this type of architecture the circuit overheads can be effectively minimized.

To give a better description about the practicability of those functional blocks, the design and implementation will be presented in *Section 5*, based on a practical receiver design that will be addressed in *Section 3*. In the meantime, the advantages obtained by this technique will be highlighted first.

Simplified channel up/downconversion at RF – it is due to the two-step channel selection. As described in Figure 3-4, it is based on a series of Bluetooth GFSK modulated channels that are being selected in the ISM band. To demonstrate the benefits of the proposed channel selection technique, three cases are considered and compared. Conventionally, in low-IF architecture, when the IF equals to 0.5 channel-spacing (*case 1*), each channel requires a local oscillator for downconversion. Differently, with the novel proposed architecture (*case 2*), the step-size of the frequency synthesizer is doubled, which implies that the division ratio (also named as modulus) in the PLL will be halved since selection between C_5 and C_6, or C_7 and C_8 will be done at the IF. If an integer-N PLL frequency synthesizer is utilized, the reference frequency, that must be equal to the step-size, can be accordingly doubled to shorten the PLL settling time by 50%, and to enlarge the loop bandwidth of the PLL for suppressing the in-locked oscillator-phase noise. Moreover, higher reference frequency also increases the PLL damping ratio, thus improves the PLL stability. Furthermore, since the division ratio is proportional to the close-in phase noise of the PLL, halving that ratio also reduces the power of the close-in phase noise by a factor of 4 and simplifies the frequency divider anatomy because half of the locking positions are saved. The detailed relationships among all of these issues have been clearly analyzed and discussed in [3.9].

The principles above referred can be further extended to other or multiple IF values. For instance, if both 0.5 and 1.5 channel-bandwidth spacings are considered (i.e., double IFs), the frequency synthesizer locking positions could then be changed to *case 3*, as shown in Figure 3-4 already. The channels C_9, C_{10}, C_{11} and C_{12} would be downconverted together by the RF local oscillator, $f_{LO,9\text{–}12}$. The selection between the four channels can be performed by a filter with four center-frequency positions, and a two-frequency-four-phased IF I/Q-LO. The step-size of the FS can be extended to four channels spacing to further enhance the abovementioned relaxations. These features are decisively important for frequency-hopping spread-spectrum (FHSS) systems such as Bluetooth, where the PLL settling time and phase noise always contribute as a design trade-off. RF-to-IF channel partitioning is therefore a key issue to establish a compromise between these two relevant PLL characteristics.

Similarly, the abovementioned considerations are also valid for a two-step-up transmitter; allowing only one specification-relaxed FS and local oscillator to fulfill both transmit and receive operations in a transceiver.

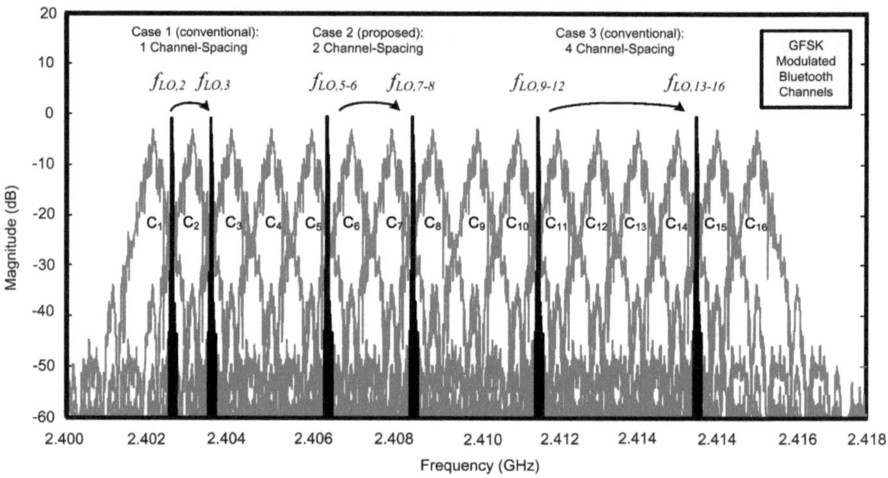

Figure 3-4. LO locking positions with conventional structures (*case 1*) and with the novel proposed 2-SCS technique (*cases 2 and 3*)

Facilitated reconfigurations in receiver and transmitter – another corollary of two-step channel selection is the fact that when two (or four) narrowband channels are considered as a wideband channel (in Figure 3-4, *cases 2 and 3*), the local oscillator can now be located at the carrier of the wideband channel to perform direct frequency down(up)conversion, i.e., a zero-IF (direct-up) operation! Of course, the bandwidth (BW) of two (or four) narrowband channels is usually not exactly equal to a wideband channel, but in order to fix it the cost would be a BW-tunable LPF, which is commonly needed in multistandard compatible designs [3.10]. The IF-to-baseband downconversion in low-IF receiver can be bypassed, resulting in a clear advantage that the receiver can operate in low-IF mode for narrowband, or zero-IF mode for wideband. Similarly, reusing this principle in the transmitter offers two operating modes, i.e., two-step-up and direct-up.

To show how these two advantages alleviate multistandard receiver design, an example reinforcing these techniques will be presented next. For simplicity, two narrowband channels in low-IF mode would be considered as one wideband channel in zero-IF one.

3. LOW-IF/ZERO-IF RECONFIGURABLE RECEIVER DESIGN

3.1 System-design overview

The three selected standards for our demonstration are Bluetooth, IEEE 802.11FH and HomeRF. Their key physical layer (PHY) specifications are summarized in Table 3-1. As the table shows, they are highly analogous to each other and operate in the same frequency band, and only one antenna and transceiver RF AFE will be sufficient to satisfy both transmit and receive operation requirements, allowing the implementation of compact and high-performance multistandard transceivers. Moreover, in the baseband AFE, automatic-gain control (AGC) blocks cooperating with both filters and A/D converters will allow efficient equalization (likely required in 802.11FH and HomeRF), as well as coherent Gaussian frequency shift-keying (GFSK) demodulation in the digital signal processor (DSP). The remaining design challenges associated with their different channel spacings and channel bandwidths lead to several design trade-offs or extra requirements on the functional blocks, which can be highlighted as follows.

Table 3-1. Summary of key PHY specifications of 2.4-GHz ISM band FHSS standards

	Channel Spacing	Data Rate	Distance	Hop Rate
Bluetooth	1 MHz	1 Mbps	10 M	1,600 Hop/s
HomeRF	5 MHz	5,10 Mbps	100 M	75 Hop/s
IEEE802.11FH	1 MHz	1,2 Mbps	150 M	2.5 Hop/s

Interference by low-frequency disturbance or image – as mentioned before, the image problem can be considerably relaxed if the receiver is implemented in zero-IF being also highly appropriate for the wideband HomeRF. However, $1/f$-noise and dc-offset interferences require ac-coupling or other offset-cancellation circuits, which can damage the narrow-band

Bluetooth and IEEE 802.11FH signals while lengthening the receiver settling time. Alternatively, employing the low-IF architecture to avoid the low-frequency disturbance, it would be at the circuit overhead of an IF downconverter and an IF I/Q-LO, a RC–CR network or a clock-frequency divider for I/Q-phase generation, as well as of an increase in the required image rejection.

Frequency synthesizer and local oscillator – although the covered standards operate in the same spectrum, this does not eliminate the need for different locking positions and step-sizes for different channel spacings. Moreover, due to the frequency-hopping spread-spectrum feature of covered standards, agile frequency synthesizer and high spectral-purity oscillator are still mandatory.

Channel-selection filter – independent to the architecture (low-IF or zero-IF), a filter with tunable bandwidth is required to comply with both narrowband and wideband channels.

3.2 Proposed receiver architecture

A low-IF/zero-IF reconfigurable receiver, as depicted in Figure 3-5, is proposed to address the aforesaid trade-offs and requirements. The receiver, which would be in low-IF mode for Bluetooth and IEEE 802.11FH, and in zero-IF mode for HomeRF, will be composed by the following blocks:

IF selection for different applications – the standard approval selectivity requirement is the main concern for IF selection. For Bluetooth, the power of the first adjacent channel is comparable to the desired channel as shown in Figure 3-6. To stay far from the low-frequency disturbance and to compromise with the image rejection, a half-channel-spacing IF is chosen. In case the low-frequency disturbance is still serious, higher IF values can be selected, of course, the overhead is the highly increased image-rejection requirement. For these particular applications, an IF equals to one-half of the CS is selected to match the proposed channel-selection technique on one hand, and to minimize the low-frequency disturbance and the required image rejection (~20 dB) on the other. The situation is similar for IEEE 802.11FH. However, for HomeRF, its bandwidth is much wider and the frequency-hopping speed is slow, zero IF with dc-offset cancellation is chosen.

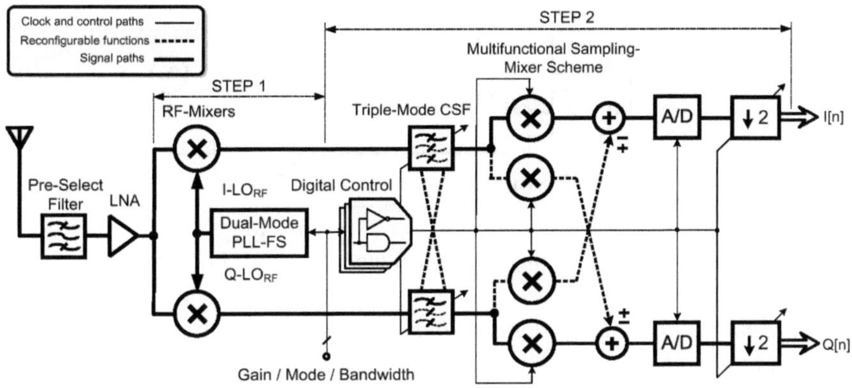

Figure 3-5. Proposed low-IF/zero-IF reconfigurable receiver

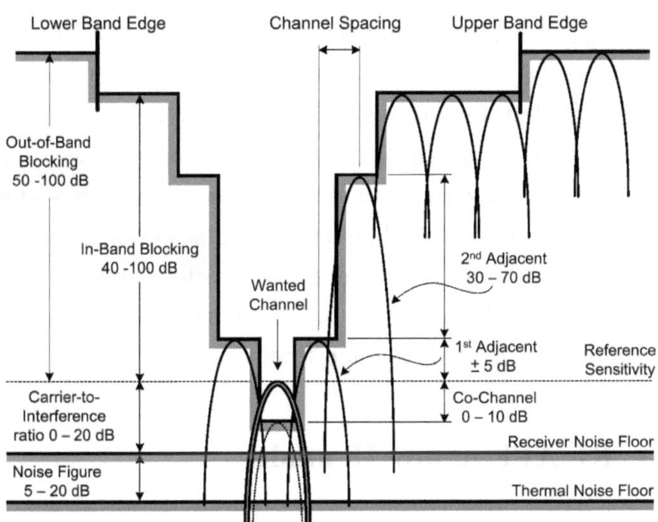

Figure 3-6. Selectivity requirement of Bluetooth

Dual-mode FS – Since the RF FS only provides a coarse channel selection, both phase-noise and settling-time requirements are relaxed. This implies the use of a simple integer-N PLL, rather than its fractional-N counterpart that suffers from spurs. Moreover, in integer-N architecture, only two additional frequency dividers would be sufficient to provide two different step-sizes based on one reference clock, the reference frequency therefore can be kept high enough to maintain low phase noise. A possible topology for the purposed receiver is illustrated in Figure 3-7.

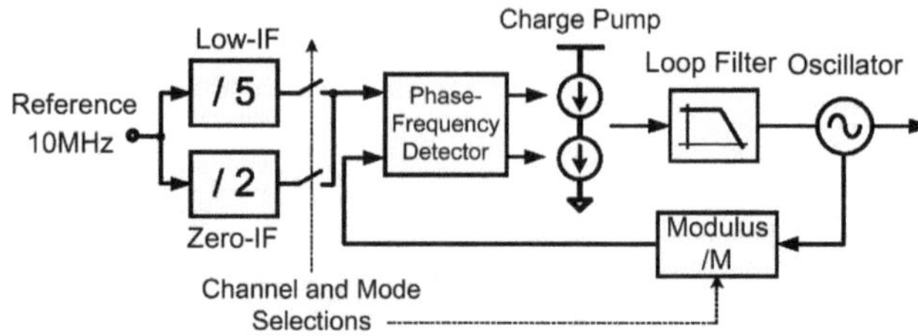

Figure 3-7. Single-band dual-mode integer-*N* PLL-FS

Triple-mode channel-selection filter – to achieve high dynamic range and also provide image rejection, active-*RC*-based polyphase filter is usually the most adequate architecture, which also simultaneously allows programmability in gain, bandwidth and center frequency. Additionally, by using some digital control, the filter can be configured as a wideband LPF for zero-IF, or a narrow-band positive-/negative-complex bandpass filter (+C-BPF/-C-BPF) for low-IF.

Multifunctional sampling-mixer scheme – in low-IF mode, the scheme performs double-quadrature sampling for downconversion (i.e., I/Q phases are inherently generated digitally), channel selection and secondary image rejection. Alternatively in zero-IF mode, the scheme executes simple baseband sampling for digitization [3.11]. The mode selection is obtained again by simple digital control, and is embedded in the clock-phase generator. The detailed design and implementation are addressed later; meanwhile, the overall working principles are presented first.

3.3 Step-1: RF AFE in low-IF mode

The reconfigurable receiver architecture (Figure 3-5) is characterized by performing the channel selection in two steps. Similar to the conventional approach in the RF AFE, the first step comprises a RF selection that will include a preselection filter and a low-noise amplifier (LNA) for acquiring and amplifying the required RF channels. After that, the FS performs the downconversion at the first stage of mixers. The corresponding CSSF illustrations are shown in Figure 3-8(a), with the selected IF value set to 0.5

channel-spacing and the assumption that the sought radio channels, at the first step of 2-SCS, are channels C_N and C_{N+1}, leading to the preselected RF input signal $x_{RF}(t)$,

$$x_{RF}(t) = C_N \cos[2\pi f_N t + \Phi_N(t)] + C_{N+1} \cos[2\pi f_{N+1} t + \Phi_{N+1}(t)] + C_{N+2} \cdots, \qquad (3.1)$$

where C_N, f and Φ are the envelope, frequency and phase of the radio channels, respectively. In the second step, depending on the desired channel, which will be either C_N or C_{N+1}, two alternatives should be considered.

3.4 Step-2: IF AFE in low-IF modes A and B

Illustrated in Figure 3-8(b) and (d) are the proposed second steps in low-IF modes A and B, respectively. The channel-selection filter would be implemented by a reconfigurable complex bandpass filter with two possible passbands ("dual-mode") that can be centered at either $-f_{IF}$ or f_{IF} for selecting between channels C_N and C_{N+1}, respectively. The programmable A-DQS, performed by a multifunctional sampling mixer, will result in either a backward (BS) or forward (FS) frequency shifting in low-IF modes A or B, respectively. Once the sampling frequency is four times of the IF (i.e., 2 MHz in this case), [3.12][3.13], this kind of sampling technique can inherently provide downconversion and sampling functions, with an additional flexibility for controlling the acquisition of either the upper or lower sideband. In forward frequency shifting, channel C_N will be obtained at $\pm nf_s$ for n=0, 1, 2..., whereas channel C_{N+1} (image of C_N) is shifted to $\pm nf_s/2$ for n=1, 3, 5.... Similar results can also be obtained in backward frequency shifting, where the roles of channels C_N and C_{N+1} will be reversed. Both cases do not suffer from the problematic dc offset and $1/f$ noise as the frequency values are now shifted to $\pm nf_s/4$, for n=1, 3, 5.... Afterwards, the $y(nT)$ can be digitized by two Nyquist A/Ds, and the final I and Q data at a rate of $f_s/2$ can be obtained after passing through a high-order digital decimation filter for sharp-transition filtration and decimation.

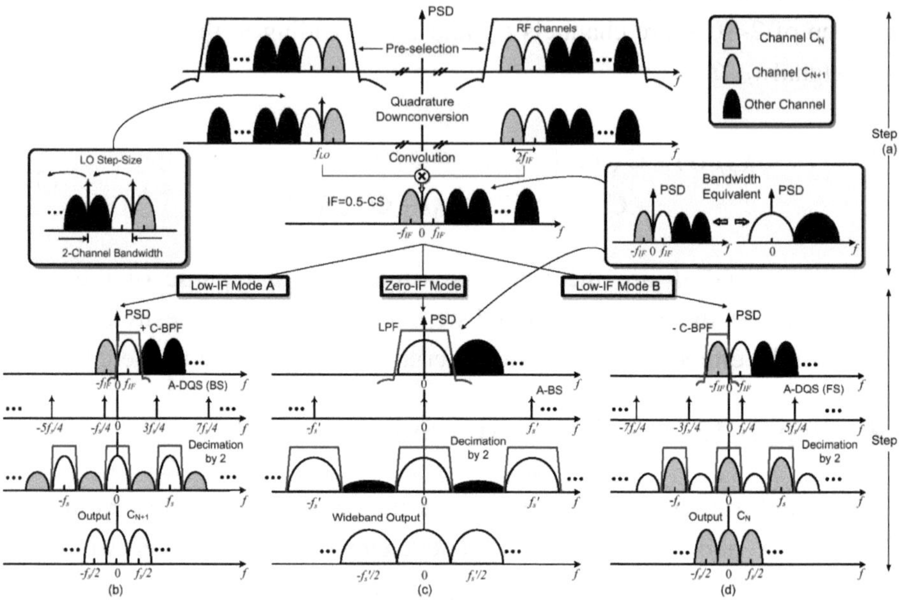

Figure 3-8. CSSF illustrations of receiver showed in Figure 3-5: (a) step-1: RF AFE (b) step-2: low-IF mode-A (c) step-2: zero-IF (d) step-2: low-IF mode-B

3.5 Step-2: IF AFE in zero-IF mode

With the same RF AFE, the main difference in the zero-IF mode, when compared with low-IF, relies on the baseband part. Once two narrowband channels are considered as a single wideband, as shown in the upper part of Figure 3-8(c), the local oscillator is at the same frequency as the channels' carrier in order to directly downconvert it to dc. The channel-selection filter should be now in lowpass mode. Regarding the double-quadrature sampling, it will be replaced by its subset operation, i.e., analog-baseband sampling (A-BS). The sampling frequency f_s is increased to 10 MHz in order to match the HomeRF standard with an oversampling ratio (OSR) of two to reduce aliasing. Clearly, independent to the modes of operation, the final demodulation is performed at dc again, allowing the use of high-effective analog zero-IF GFSK demodulator [3.14] or digital coherent demodulator.

The subsequent digital operations in zero-IF mode will be analogous to those in low-IF mode, and they can be implemented in the DSP.

4. DIRECT-UP/TWO-STEP-UP RECONFIGURABLE TRANSMITTER DESIGN

The proposed technique implemented in the receiver can be analogously adopted in the transmitter to allow one PLL frequency synthesizer per transceiver and it shares the same phase-noise and settling-time relaxations. As depicted in Figure 3-9, a direct-up/two-step-up transmitter architecture is presented. In direct-up mode, single upconversion has a simple and efficient architecture, but it suffers from LO pulling as the frequency of the LO is the same as the desired frequency location. In two-step-up mode, the first IF is set to half channel spacing to reduce the effect of LO-LO self-modulation at RF since the LO frequency can be offsetted from the RF signal frequency. Regarding the IF value, again, it can be selected as either a positive or negative IF, like the receiver path, through the developed multifunctional sampling-mixer scheme. Different from the conventional two-step-up transmitter [3.15], the A-DQS upconversion can be embedded into the S/H units of typical switched-capacitor digital-to-analog (D/A) converters, or implemented in current mode for current-steering D/A converters [3.16]. The LPF used in conventional case can be replaced by the proposed triple-mode filter which rejects, simultaneously, the residue images due to sample-and-hold effects and I/Q mismatch. Moreover, with capacitive coupling embedded in the cascaded stages, $1/f$ noise and dc offset generated from the filters can be suppressed without degrading the signal quality, because the signal has been now

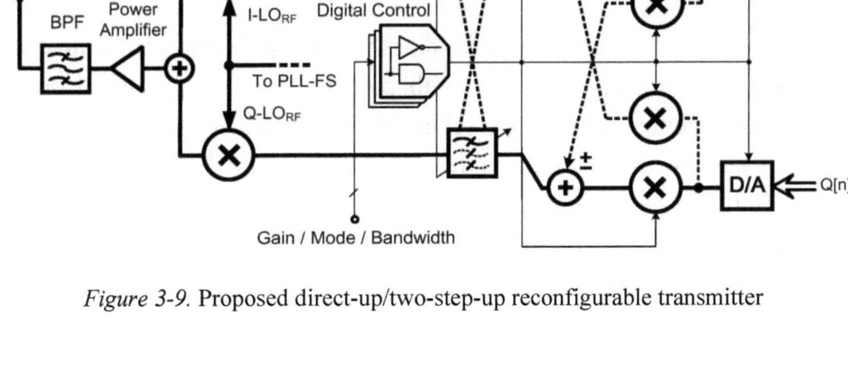

Figure 3-9. Proposed direct-up/two-step-up reconfigurable transmitter

frequency shifted to the IF prior to filtering. The final upconversion is carried in the second step.

5. RECONFIGURABLE IF AFE DESIGN

To implement a receiver adequate for the previously referred applications (except for the simple modification needed in the FS presented in Figure 3-7), only the IF-to-baseband functional blocks need to be reconfigured as introduced next.

5.1 Triple-mode channel-selection filter

Basic principles – to fulfill the filtering modes described in Figure 3-8, only a small number of center-frequency locations is necessary. By doing lowpass-to-complex-bandpass linear transformation defined by $j\omega \rightarrow j\omega - j\omega_c$ (where ω_c is the tunable center frequency), the characteristics of the filter such as phase linearity and passband ripple are all preserved after transformation. If the filter is an even *n*th-order all-pole filter, the transfer function can be written as,

$$H(s + j\omega_c) = \prod_{i=1}^{n/2} \frac{K_i \cdot \omega_{p,i}^2}{(s + j\omega_c)^2 + (\omega_{p,i} / Q_{p,i})(s + j\omega_c) + \omega_{p,i}^2}, \qquad (3.2)$$

whereas an odd *n*th-order one has the form,

$$H(s + j\omega_c) = \frac{K_r \cdot \omega_r}{(s + j\omega_c) + \omega_r} \cdot \prod_{i=1}^{(n-1)/2} \frac{K_i \cdot \omega_{p,i}^2}{(s + j\omega_c)^2 + (\omega_{p,i} / Q_{p,i})(s + j\omega_c) + \omega_{p,i}^2}, \qquad (3.3)$$

where $s=j\omega$, K_i and K_r are constants, ω_r is the real pole and $Q_{p,i}$ is the quality factor of the conjugate pole pairs with pole frequency $\omega_{p,i}$. Based on (3.2) and (3.3), the generalized block schematics of uniquad and biquad implementation can be depicted in Figure 3-10. It shows that by changing the quadrature-feedback factors between the I and Q channels, the center frequency can be controlled to form the intended filtering modes, i.e., lowpass ($\omega_c = 0$), negative ($\omega_c = -\omega_c$) or positive ($\omega_c = +\omega_c$) complex bandpass.

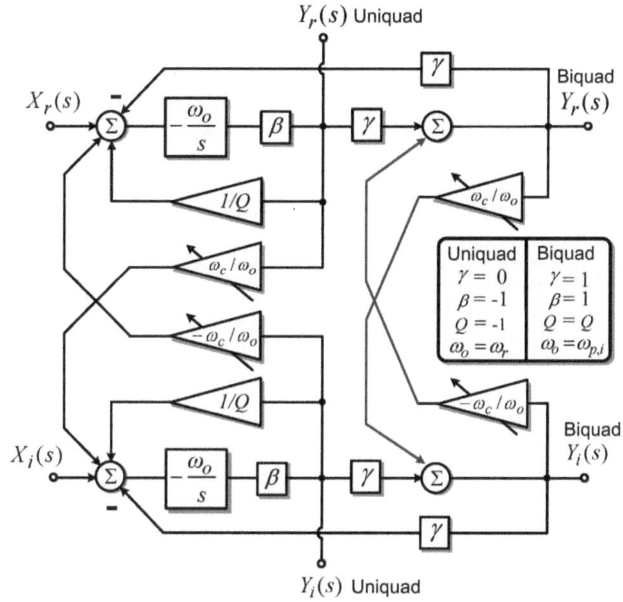

Figure 3-10. Block schematic of a center-frequency-tunable uniquad/biquad stage

Implementation – wireless transceivers require a high-linear filter for channel selection, an active-*RC* realization constitutes an adequate choice. Figure 3-11 shows the simplified fifth-order implementation, where the complex uniquad has been obtained by modifying the circuit presented in [3.17] (which has been traditionally used to construct high-order polyphase filters through cascading) with the introduction of a center-frequency controller. However, to realize a higher-order Butterworth characteristic with complex poles, a novel complex biquad structure is employed here, which is based on the Tow-Thomas architecture. By using a simple capacitor (resistor) bank [3.18] in the feedback capacitor C_{fb}s (forward resistor R_fs), the bandwidth (gain) can be digitally scaled. The filter center frequency will be set up by a 2-bit digital controller, which organizes the MOS array of switches in such a way that it will either connect the I/Q cross-feedback resistors R_{qf}s to the common-mode voltage for centering the filter at dc, or switches the differential terminals to centre the filter at either $+\omega_c$ or $-\omega_c$. This type of arrangement allows the elimination of the loading effects in mode-switching (mode-A or mode-B), and reduces the distortion caused by the MOS switches since R_{qf} is always connected to either the common mode or the virtual ground of the OpAmp. In addition, R_{qf} is implemented by a resistor bank to further enlarge the center-frequency adjustability.

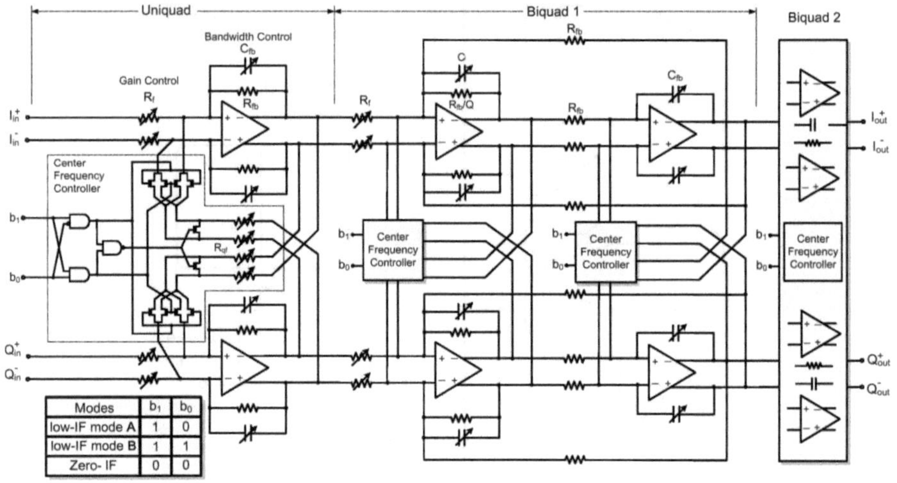

Figure 3-11. Triple-mode CSF with active-*RC* realization

Another important consideration is the fact that ac coupling does not raise any problem in both low-IF and zero-IF modes because HomeRF is wide-band and the corresponding settling time requirements are very low (i.e., slow-frequency hopping, as presented in Table 3-1). For the Bluetooth and IEEE 802.11FH, since 99% of their power is concentrated in the frequency range between 70 kHz and 430 kHz, this implies that either ac coupling or a slight increase in the dynamic range of the A/D converter can effectively alleviate the dc-offset and $1/f$-noise problems.

Table 3-2. Pole positions of the triple-mode filter

Poles	Lowpass Prototype (rad/s)	Lowpass at DC (MHz)	Complex bandpass at +0.5MHz (MHz)	Complex bandpass at -0.5MHz (MHz)	Complex bandpass at +1.5MHz (MHz)	Complex bandpass at -1.5MHz (MHz)
BW	1	(LP) 2.5	(BP) 1	(BP) 1	(BP) 1	(BP) 1
P_1	−0.309 + 0.951j	−0.773 + 2.378j	−0.155 + 0.976j	−0.155 − 0.025j	−0.155 + 1.976j	−0.155 − 1.025j
P_2	−0.309 − 0.951j	−0.773 − 2.378j	−0.155 + 0.025j	−0.155 − 0.976j	−0.155 + 1.025j	−0.155 − 1.976j
P_3	−0.809 + 0.588j	−2.023 + 1.470j	−0.405 + 0.794j	−0.405 − 0.206j	−0.405 + 1.794j	−0.405 − 1.206j
P_4	−0.809 − 0.588j	−2.023 − 1.470j	−0.405 + 0.206j	−0.405 − 0.794j	−0.405 + 1.206j	−0.405 − 1.794j
P_5	−1	−2.5	−0.500 + 0.500j	−0.500 − 0.500j	−0.500 + 1.500j	−0.500 − 1.500j
	Poles Q-factor: 0.500-1.618		Poles Q-factor: 0.506-3.188		Poles Q-factor: 1.571-6.394	

Simulation results – the pole positions of the triple-mode filter is listed in Table 3-2, and pictorially presented in Figure 3-12. The simulated frequency responses of the baseband filter, obtained with the parameters of a 0.35-μm CMOS in SPECTRE, are shown in Figure 3-13. The OpAmp employs a single-pole model exhibiting 80-dB dc gain, 993-MHz gain-bandwidth product and with an input parasitic capacitance of 0.5 pF. In zero-IF mode

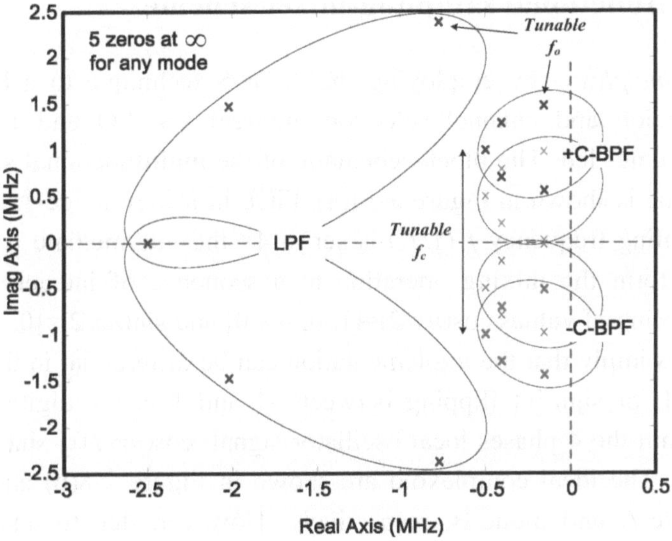

Figure 3-12. Pole-zero plot of the triple-mode filter

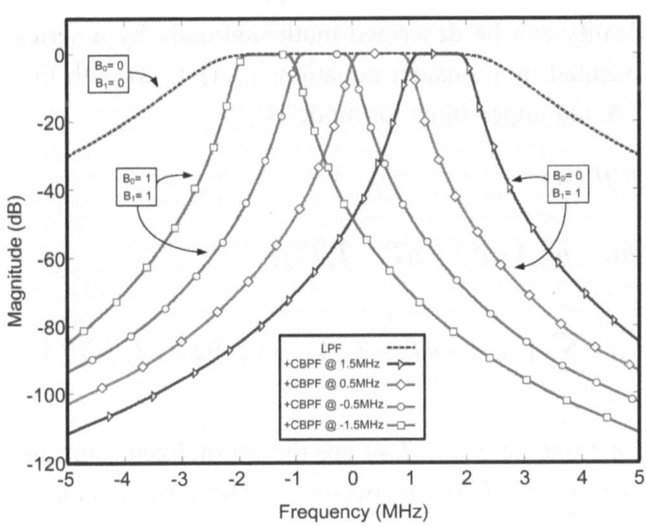

Figure 3-13. Magnitude response of the triple-mode filter

with 5-MHz BW, the first adjacent channel suppression is approximately 30 dB. In low-IF mode, the BW is changed to 1 MHz, and the center frequency is controlled to be at either 0.5 or 1.5 MHz to show the possibility of double-IF channel selection, being the first adjacent channel rejection also approximately 30 dB.

5.2 Multifunctional sampling-mixer scheme

Basic principles – by employing the A-DQS technique to achieve the downconversion and channel selection, no real FS, LO and I/Q phase generator are needed. The block schematic of the multifunctional sampling-mixer scheme is shown in Figure 3-14(a). First, in low-IF mode-A or mode-B, the sampling frequency f_s $(1/T_s)$ is set to 4× the intermediate frequency, f_{IF}, to transform the mixing operation to a sequence of integer-weighted analog sampling of values $\cos(n\pi/2)$=[1, 0, –1, 0] and $\sin(n\pi/2)$=[0, 1, 0, –1]. These values imply that the implementation can be differential in the analog domain [3.4] or sign-bit flipping between –1 and 1 in the digital domain [3.5] to obtain the 4-phases local oscillator signal: $\cos(n\pi/2)\pm j\sin(n\pi/2)$ for n=1, 2, 3… The ideal complexoid are shown in Figure 3-8(b) and (d) for low-IF mode-A and mode-B, respectively. However, due to unavoidable gain and I/Q-time-skew mismatches between the I and Q channels, the frequency-shifting and image-rejection features of A-DQS will not be exact. Such a nonideality can be described mathematically by a series of impulse samples represented by (consider equations (3.4)–(3.8) with lower signs for low-IF mode-A and upper signs for mode-B),

$$\begin{aligned} P(t) &= P_I(t) + jP_Q(t) \\ &= \sum_{n=-\infty}^{\infty} [\delta(t-nT_s) - \delta(t-nT_s-T_s/2)] \\ &\quad + j(1+\alpha)\sum_{n=-\infty}^{\infty} [\pm\delta(t-nT_s-T/4-\sigma) \mp \delta(t-nT_s+T/4-\sigma)] \end{aligned} \quad , \quad (3.4)$$

where α and σ (assume $\sigma << T/4$) are the normalized gain and time-skew error between $P_I(t)$ and $P_Q(t)$, respectively. The time-domain illustration is shown in Figure 3-14(b) for low-IF mode-A, where $x_I(t)=V_p\cos(\omega_{IF}t)$ and

$x_Q(t)=V_p\sin(\omega_{IF}t)$ are assumed as the differential inputs, and only $y_I(nT)$ is presented for simplicity. The Fourier transform of $P(t)$ yields,

$$\begin{aligned} P(j\omega) &= \sum_{k=-\infty}^{\infty} 2\pi a_k \delta(\omega-\omega_{IF}) + j\sum_{k=-\infty}^{\infty} 2\pi b_k \delta(\omega-\omega_{IF}) \\ &= \sum_{k=-\infty}^{\infty} 2\pi c_k \delta(\omega-\omega_{IF}) \end{aligned} , \quad (3.5)$$

where $\omega_s=2\pi f_s$, a_k, b_k and c_k are the Fourier coefficients of the real part, imaginary part, and their complex-sum, respectively, given by,

$$a_k = \begin{cases} \dfrac{2}{T} & \text{for } k=2n+1 \\ 0 & \text{otherwise} \end{cases} , \quad (3.6)$$

$$b_k = \begin{cases} \dfrac{2}{T}(1+\alpha)e^{-jk\pi\left(\frac{2\sigma}{T}\pm\frac{1}{2}\right)} & \text{for } k=2n+1 \\ 0 & \text{otherwise} \end{cases} , \quad (3.7)$$

and

$$c_k = a_k + jb_k = \begin{cases} \dfrac{2}{T}\left[1+(1+\alpha)e^{-jk\pi\left(\frac{2\sigma}{T}\pm\frac{1}{2}\right)}\right] & \text{for } k=2n+1 \\ 0 & \text{otherwise} \end{cases} , \quad (3.8)$$

for $n=\pm1, \pm2, \pm3\ldots$. The nonideal complexiod output spectrums, $P_I(j\omega)$, $P_Q(j\omega)$ and $P_I(j\omega)+jP_Q(j\omega)$ in low-IF mode-A are shown in Figure 3-14(c), and the image-rejection ratio (IRR) can be quantified by,

$$IRR = \frac{|c_1|^2}{|c_{-1}|^2} = \frac{1+(1+\alpha)^2+2(1+\alpha)\cos(\varepsilon)}{1+(1+\alpha)^2-2(1+\alpha)\cos(\varepsilon)} , \quad (3.9)$$

where ε is the time-skew-induced phase mismatch such that $\varepsilon=2\pi\sigma/T$. From (3.6), the IRRs are 32 dB (34 dB) for a 0.025 (0.02) relative gain mismatch when combined with 2.5° (2°) phase mismatch. These values of IRRs are practically achievable and they can be added together with the image rejection provided by the triple-mode channel-selection filter providing in total ~60-dB image rejection in the IF AFE, which represents a significant improvement when compared with conventional structures that achieve usually 30 dB without tuning or trimming [3.19].

It is also noteworthy that the proposed sampling-mixer scheme is different from the traditional subsampling mixer because the sampling frequency is Nyquist and there is prefiltering prior to the sampling. The signal-band noise floor therefore will not be increased by any subsampling factor due to wideband-noise aliasing [3.20].

In zero-IF mode, the sampling mixer is reduced to a subset of the previous structure, and it will be designated as A-BS also shown in Figure 3-14(a). It can be obtained by deactivating the quadrature-phase sampler $P_Q(t)$, and changing $P_I(t)$ and $P_I(j\omega)$ to $P'_I(t)$ and $P'_I(j\omega)$, respectively, which can be expressed by,

$$P'_I(t) = \sum_{n=-\infty}^{\infty} \delta(t - nT'_s), \tag{3.10}$$

$$P'_I(j\omega) = \sum_{k=-\infty}^{\infty} 2\pi a'_k \, \delta(\omega - k\omega'_k), \tag{3.11}$$

where

$$a'_k = \left\{ \frac{1}{T} \text{ for } k = 0, \pm 1, \pm 2, \ldots \right. . \tag{3.12}$$

Implementation – the circuit solution implementing those sampling schemes is depicted in Figure 3-15(a), which is based on a simple half-delay offset-compensated sample-and-hold (S/H) pair, which can serve as the front-end of an A/D [3.21]. In low-IF mode, the sequential clock sampling 1–4 performs

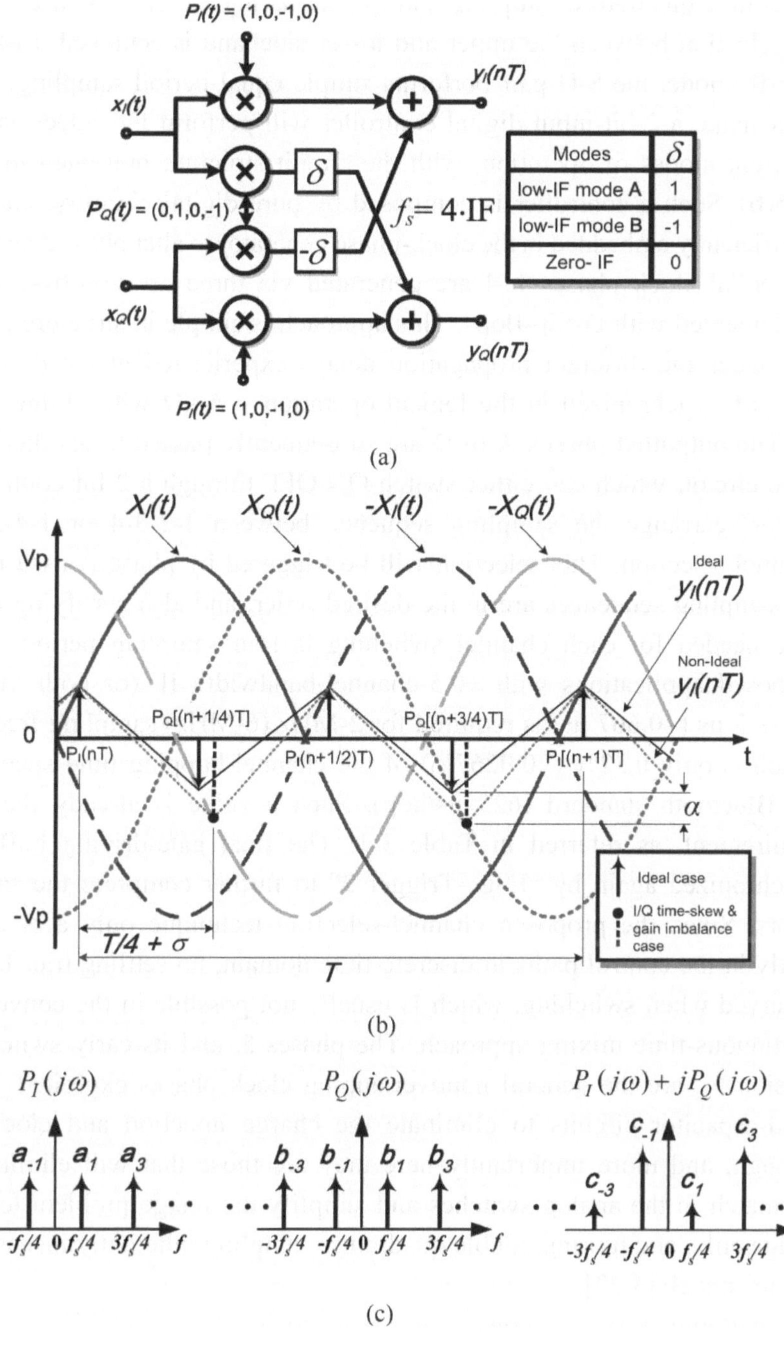

Figure 3-14. (a) Block schematic of A-DQS/A-BS-reconfigurable scheme (b) Time-domain illustration of ideal and nonideal A-DQS in low-IF mode A (c) Nonideal output spectrum of A-DQS in low-IF mode A

the double quadrature sampling and by alternating the clock phases 2 and 4 the selection between the upper and lower sideband is achieved. Instead, in zero-IF mode, the S/H pair performs simple equal-period sampling. On the other hand, a 2-bit-input digital controller will perform the selection of the different modes of operation, with the circuit structure presented in Figure 3-15(b). Such a controller is composed by pure digital circuitry, and it can be efficiently embedded in the clock-phase generator to alter phase 2 and 4. The sequential clock phases 1–4 are generated via three division-by-2 circuits implemented with D-flip-flops. This approach is simple in structure and also eliminates the different propagation delays experienced in the dividers as they are synchronized in the logical operation – AND with "Edge-Trigger 1." The outputted phases A to D are subsequently passed to another simple logic circuit, which can either switch-ON/OFF through a 2-bit control code and/or rearrange the sampling sequence between 1-2-3-4 or 1-4-3-2 for channel selection. This selection will be triggered by phase A assuring that the sampling sequences are in the desired order, and also certifying that the time needed for each channel switching is four sampling periods. In the proposed applications with ±0.5-channel-bandwidth IF (or both ±0.5 and ±1.5), 2 μs (~0.667 μs) is required for 2-MHz (6-MHz) sampling frequency, which is only 0.32% (~0.0267%) of the channel-hopping time specified in the Bluetooth standard (625 μs/hop). Such a value is already the fastest requirement, as referred in Table 3-1. The final gate-driving buffers are synchronized again by "Edge-Trigger 2" to further compress the sampling errors. Since the proposed channel-selection technique only acts transparently in the control paths in discrete-time domain, no settling transients are observed when switching, which is usually not possible in the conventional continuous-time mixing approach. The phases 5, and its early switched-off version 5', are the general nonoverlapping clock phases exploited in switched-capacitor circuits to eliminate the charge injection and clock feed-through, and more importantly here they are those that will eliminate the mismatch in the analog switches and simplify the image problem to a self-image only, and turning double quadrature sampling inherently insensitive to I/Q mismatch [3.22].

Simulation results – considering that the baseband sampling is a well-known technique, its simulation results will be omitted here and then only the results of double-quadrature sampling in forward shifting will be addressed next.

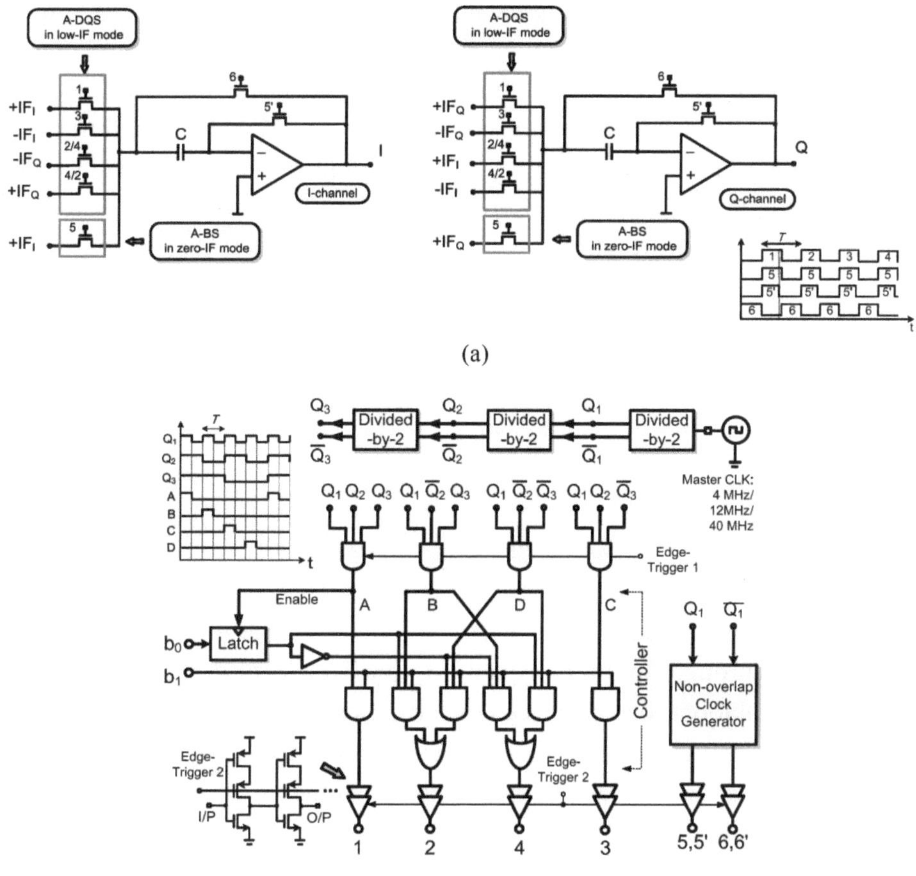

Figure 3-15. (a) A multifunctional sampling-mixer scheme and (b) Channel/mode selection embedded clock-phase generator (the digital control code b_0 and b_1 are given in Figure 3.11)

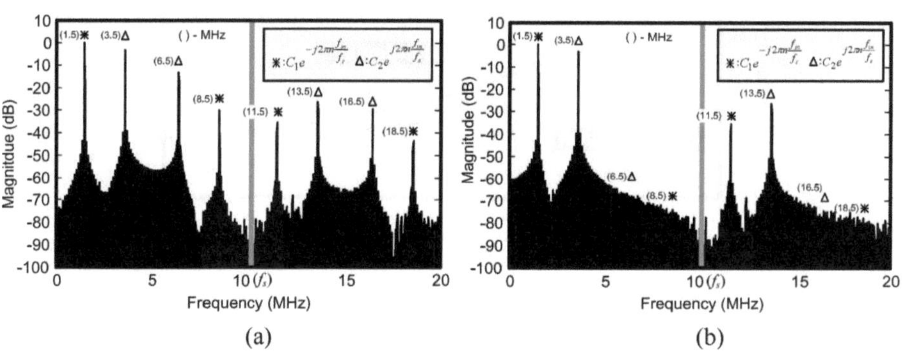

Figure 3-16. Simulated PSDs: (a) I channel and (b) |I+jQ|

By applying a pair of complex input signal, $C_1e^{-j2\pi f_{in}t} + C_2e^{j2\pi f_{in}t}$, with f_{in}=1 MHz and setting the sampling frequency, f_s, to 10 MHz, the obtained power spectrum density (PSD) of the I channel is shown in Figure 3-16(a) (the Q channel's result would be identical in magnitude but different in phase). Taking the complex sum PSD |I+jQ| will lead to the simulated results plotted in Figure 3-16(b). As observed, the input is not only sampled-and-held but also forwardly shifted by 2.5 MHz ($f_s/4$) such that the sampled components,

$$C_1e^{-j2\pi n\frac{f_{in}}{f_s}} \text{ and } C_2e^{-j2\pi n\frac{f_{in}}{f_s}},$$

will be located at

$$nf_s + \frac{f_s}{4} + f_{in} \text{ and } nf_s + \frac{f_s}{4} - f_{in} \text{ for } n = 1, 2, 3\ldots,$$

respectively. The attenuation of their magnitudes is due to the sample-and-hold effect. Through optimum system-to-transistor-level design, the image-rejection simulations based on certain artificial mismatch assignments, on the *I* and *Q* channels, achieved IRR=41/50/76 dB for 2% gain, 0.5° phase and 2% capacitance mismatches, respectively, as shown in Figure 3-17.

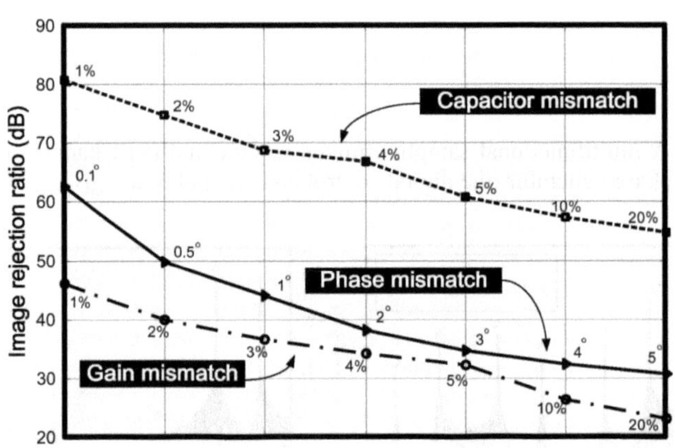

Figure 3-17. Simulated IRR versus capacitance, gain and phase mismatch between I and Q channels

Table 3-3. Comparison of the proposed receiver and transmitter architectures to the existed solutions

RECEIVERS						
Architecture	Image Susceptibility	1/*f*-Noise and DC-Offset Susceptibility	No. of Mixer	No. of Frequency Synthesizer	No. of Local Oscillator	No. of Filter and Types
Zero-IF	Low	High	2 at RF	1 at RF	1 at RF	2 LPFs
Low-IF	Medium	Low	2 at RF, 4 at Low IF	1 at RF	1 at RF, 1 at IF	1 C-BPF
Wide-band IF	Medium	Low	1 at high IF, 2 at Low IF	1 at IF	1 at RF, 1 at IF	2 LPFs
Low-IF /Zero-IF Reconfigurable (Proposed)	Low [1]	Low/High Flexible	2 at RF [2]	1 at RF [2]	1 at RF [2]	1 C-BP/LP Tunable Filter
TRANSMITTERS						
Architecture	Image Susceptibility	1/*f*-Noise and DC-Offset Susceptibility	No. of Mixer	No. of Frequency Synthesizer	No. of Local Oscillator	No. of Filter and Types
Direct-up	Medium	High	2 at RF	1 at RF	1 at RF	2 LPFs
Two-step-up	Medium	Low	2 at RF, 4 at Low IF	1 at RF	1 at RF, 1 at IF	2 LPFs
Direct-up /Two-step-up Reconfigurable (Proposed)	Medium/Low [4] Flexible	High/Low Flexible	2 at RF [2]	1 at RF [2]	1 at RF [2]	1 C-BP/LP Tunable Filter

[1] In low-IF mode, the proposed IF AFE rejects the image twice, by the sampling-mixer scheme and the triple-mode CSF

[2] The IF channel selection and downconversion are embedded functions of an S/H pair. No real mixer, oscillator and synthesizer are needed at IF

[3] I/Q imbalances in the D/A, LPF and mixer-induced image, plus the 1/*f* noise and DC offset

[4] Embedding frequency upconversion in the D/A and replaces the LPF by a triple-mode C-BPF presenting a good compromise in interference susceptibility

6. SUMMARY

As the multistandardability of wireless terminals becomes much more imperative than before, a stand-alone operation of either low-IF or zero-IF receiver may not be adequate enough to support narrowband-wideband-mixed multistandard applications.

This chapter has introduced a *two-step channel-selection* technique to combine the beneficial features of low-IF and zero-IF downconversion in one reconfigurable receiver. As a result, with some simple alterations introduced in the RF AFE (two frequency dividers added to the FS), both narrowband and wideband signals can be processed flexibly in their respectively preferred low-IF and zero-IF operating modes. Additionally, the technique also relaxes the FS and LO design difficulties through channel-selection partitioning between the RF and IF AFEs, and balance the design trade-offs between image-rejection and low-frequency interference eliminations. Those techniques and functional blocks are also transformable to a reconfigurable transmitter design to form an equally flexible direct-up/two-step-up transmitter.

A comparison is made in Table 3-3 to illustrate the versatility of the developed receiver and transmitter against the conventional, demonstrating the feasibility of the proposed architectures in multiple aspects.

REFERENCES

[3.1] Leon W. Couch II, *Digital and Analog Communication Systems*, Prentice-Hall, 1987.

[3.2] J. C. Rudell, et al., "A 1.9-GHz Wide-Band IF Double Conversion CMOS Receiver for Cordless Telephone Applications," *IEEE Journal of Solid-State Circuits (JSSC)*, vol. 32, no. 12, pp. 2071–2088, Dec. 1997.

[3.3] B. Razavi, "Architectures and Circuits for RF CMOS Receivers," in *Proc. Custom Integrated Circuits Conference (CICC)*, pp. 393–400, May 1998.

[3.4] P.-I. Mak, S.-P. U and R. P. Martins, "A Novel IF Channel Selection Technique by Analog Double-Quadrature Sampling for Complex Low-IF Receivers," in *Proc. International Conference on Communication Technology (ICCT)*, vol. 2, pp. 1238–1241, Apr. 2003.

[3.5] P.-I. Mak, S.-P. U and R. P. Martins, "Two-Step Channel Selection Technique by Programmable Digital Double-Quadrature Sampling for Complex Low-IF Receivers," *IEE Electronics Letters*, vol. 39, no. 11, pp. 825–827, May 2003.

[3.6] P.-I. Mak, S.-P. U and R. P. Martins, "Two-Step Channel Selection – A Novel Technique for Reconfigurable Multistandard Transceiver Front-Ends," *IEEE Transactions on Circuits and Systems-I (TCAS-I)*, pp. 1302–1315, July 2005.

[3.7] P.-I. Mak, S.-P. U and R. P. Martins, "Two-Step Channel Selection for Wireless Receiver and Transmitter Front-Ends," *US Patent pending*, Serial No. 11/213,613, Aug. 2005.

[3.8] S. Mirabbasi and K. Martin, "Hierarchical QAM: A Spectrally Efficient DC-Free Modulation Scheme," *IEEE Communications Magazine*, pp. 140–146, Nov. 2000.

[3.9] B. Razavi, *RF Microelectronics*. Prentice-Hall, 1998.

[3.10] H. A. Alzaher, H. O. Elwan and M. Ismail, "A CMOS Highly Linear Channel-Select Filter for 3G Multistandard Integrated Wireless Receivers," *IEEE Journal of Solid-State Circuits (JSSC)*, vol. 37, pp. 27–37, Jan. 2002.

[3.11] P.-I. Mak, S.-P. U and R. P. Martins, "A Low-IF/Zero-IF Reconfigurable Receiver with Two-Step Channel Selection Technique for Multistandard Applications," in *Proc. IEEE International Symposium on Circuits and Systems (ISCAS)*, pp. 417–420, May 2004.

[3.12] P.-I. Mak, W.-I. Mok, S.-P. U and R. P. Martins, "*I/Q* Imbalance Modeling of Quadrature Transceiver Analog Front-Ends in SIMULINK," in *Proc. IEEE International Conference on Vehicular Technology (VTC)*, vol. 4, pp. 2371–2374, Oct. 2003.

[3.13] P.-I. Mak, S.-P. U and R. P. Martins, "A Front-to-Back-End Modeling of *I/Q* Mismatch Effects in a Complex-IF Receiver for Image-Rejection Enhancement," in *Proc. IEEE International Conference on Electronics, Circuits and Systems (ICECS)*, pp. 631–634, Dec. 2003.

[3.14] S. Samadian, R. Hayashi and A. A. Abidi, "Low Power Phase Quantizing Demodulators for a Zero-IF Bluetooth Receiver," *IEEE Radio Frequency Integrated Circuits Symposium (RFIC), Digest of Technical Papers*, pp. 49–52, June 2003.

[3.15] J. A. Weldon, R. S. Narayanaswami, J. C. Rudell, L. Lin, M. Otsuka, S. Dedieu, L. Tee, K.-C. Tsai, C.-W. Lee and P. R. Gray, "A 1.75-GHz Highly Integrated Narrow Band CMOS Transmitter with Harmonic-Rejection Mixers," *IEEE Journal of Solid-State Circuits (JSSC)*, vol. 36, pp. 2003–2015, Dec. 2001.

[3.16] K.-H. Ao Ieong, C.-Y. Fok, P.-I. Mak, S.-P. U and R. P. Martins, "A Frequency Up-Conversion and Two-Step Channel Selection Embedded CMOS D/A Interface," in *Proc. IEEE International Symposium on Circuits and Systems (ISCAS)*, May 2005.

[3.17] J. Crols and M. Steyaert, "An Analog Integrated Polyphase Filter for a High Performance Low-IF Receiver," *IEEE Symposium on VLSI Circuits (VLSI), Digest of Technical Papers*, pp. 87–88, June 1995.

[3.18] A. M. Durham, J.B. Hughes and W. R. White, "Circuit Architecture for High Linearity Monolithic Continuous-Time Filtering," *IEEE Transactions on Circuits and Systems-II (TCAS-II)*, vol. 39, no. 9, Sept. 1992.

[3.19] A. A. Abidi, "RF CMOS Comes of Age," *IEEE Journal of Solid-State Circuits (JSSC)*, vol. 39, no. 4, pp. 549–561, Apr. 2004.

[3.20] D. H. Shen, C.-M. Hwang, B. Lusignan and B. A. Wooley, "A 900 MHz Integrated Discrete-Time Filtering RF Front-end," *International Solid-State Circuits Conference (ISSCC), Digest of Technical Papers*, pp. 54–55, Feb. 1996.

[3.21] P.-I. Mak, C.-S. Sou, S.-P. U and R. P. Martins, "A Frequency-Downconversion and Channel-Selection A-DQS Sample-and-Hold Pair for Two-Step-Channel-Select Low-IF Receiver," in *Proc. IEEE International Conference on Electronics, Circuits and Systems (ICECS)*, pp. 479–482, Dec. 2003.

[3.22] K.-P. Pun, J. E. Franca, C. A. Leme and R. Reis, "Quadrature Sampling Schemes with Improved Image Rejection," *IEEE Transactions on Circuits and Systems-II (TCAS-II)*, vol. 50, no. 9, pp. 641–648, Sept. 2003.

Chapter 4

SYSTEM DESIGN OF A SIP RECEIVER FOR IEEE 802.11A/B/G WLAN

1. INTRODUCTION

Entered into the nanoelectronics era, SoC integration is expected to deliver substantial improvements in cost and performance for digital circuits such as microprocessor [4.1]. In sharp contrast, fabricating a mixed-signal SoC (e.g., a wireless transceiver) in nanoscale technologies is still in its infancy stage [4.2]. Operating-voltage lowering, low scalability of passives (i.e., *R*, *L*, *C*) and cosubstrate interference, are all bottlenecks that can highly obstruct such kinds of mixed-signal system from attaining high performance at low cost.

3-D stacked SiP integration [4.3], on the other hand, appears as a more promising technology. Dies fabricated with heterogeneous technologies (i.e., BiCMOS and CMOS) and optimized for each standard can be stacked as a multichip module (MCM) thought chip-to-chip interconnects, economizing on the increasingly costly mask sets required for manufacturing. Many low-cost *digital* systems such as memory [4.4] have been enabled by this technology.

This chapter tries to project the 3-D stacked SiP technology into the area of *mixed-signal* VLSI wireless systems. An IEEE 802.11a/b/g wireless-LAN transceiver, with emphasis on receiver analog baseband (BB) [4.5], is proposed. A 3D floorplan beyond die integration is described, which ensures efficient 3D routability chip-to-chip interconnects and testability.

2. SYSTEM DESIGN

2.1 Proposed 3D stacked system partition for SiP integration

It will be more comprehensive, in system level, to consider the complete 3-D stacked SiP transceiver (Figure 4-1(a)) rather than just the receiver path. Figure 4-1(b) depicts the transceiver block schematic. It is pieced by heterogeneous technologies that best-fit the layer's function. Radio-frequency (RF) signals are avoided going off-chip, except those between the antenna and the radio. The radio, depended upon the cost and performance requirements, is free from technology choices. For instance, higher sensitivity requires a lower-noise radio; BiCMOS becomes a more promising choice. For the analog baseband, the speed requirement is well affordable by the submicron CMOS, but mixed-signal options like MIM capacitors and high-resistive polysilicon resistors are preferred to minimize the silicon area. Since the covered wireless-LAN standards employs time-division multiple access (TDMA), the analog baseband not only can serve multiple standards, but also can be receiver–transmitter reconfigurable. The A/D and D/A interfaces (assuming 10-bit resolution) are resided on the digital baseband for two main reasons. First, a digital interface would require at least 40 more pins for communication between the two layers, as the receiver and transmitter chains generate I and Q signals in quadrature up/down-conversion. Second, the large switching power of the A/D and D/A interfaces can result in substantial noise on the supply rails. Both the digital baseband and MAC processor are typically integrated on the same die [4.6] and are nanoscale-process-preferred for cost and power reduction. For the power management unit, it offers four modes: standby, sleep, receive and transmit. Programmable low-dropout (LDO) regulator reduces leakage power in digital chips. One more possible layer can be a MEMS resonator, which serves as a purely on-chip timing reference.

3D stacking of multiple chips requires 3D floor plan to insure routability. The proposed 3D floor-plan can be optimized in the proposed transceiver not only in RX modes (Figure 4-2(a)) but also in the transmitter (TX) mode (Figure 4-2(b)). It comprises of three chips in each path, radio, analog BB and digital BB, for a possible full integration of the system. Without any cross bonding between chips, the analog BB chip bridges the radio to the digital BB without any complicated routing.

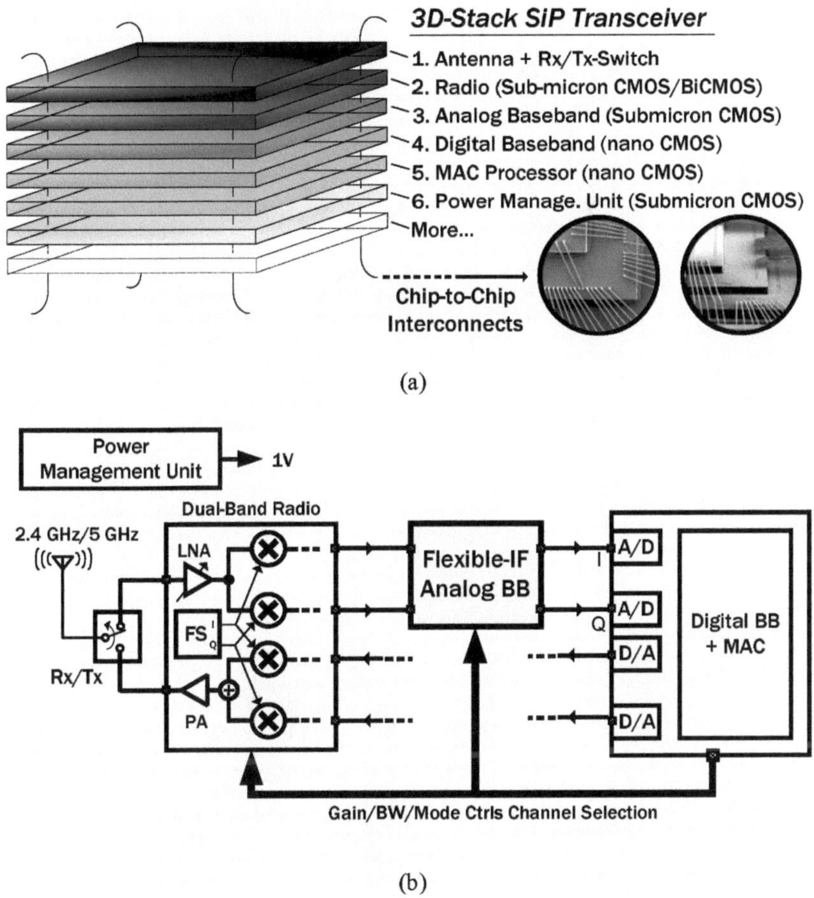

Figure 4-1. Proposed transceiver: (a) 3D stack architecture and its (b) block schematic

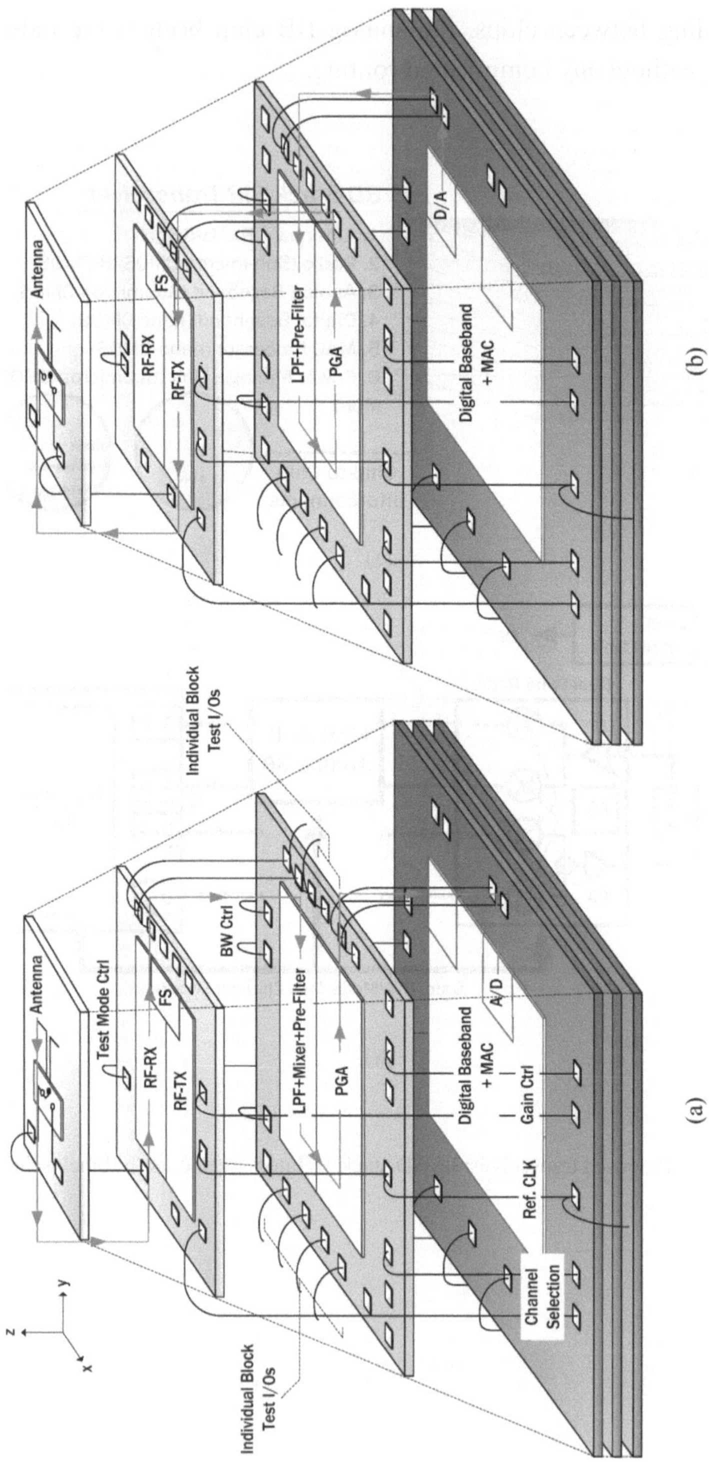

Figure 4-2. Simplified 3D floor plan and pin assignments: (a) RX and (b) TX

2.2 Proposed flexible-IF reception for multistandardability

Architecturally, receiver RF front-ends see no difference between zero-intermediate-frequency (ZIF) and low-IF (LIF) downconversion [4.7]. The analog-BB chain is therefore flexible in choosing the best-fit IF for each mode of operation. In this paper, a ZIF-LIF-mixed solution is proposed.

ZIF is well suited for 802.11b/g in complementary code keying (CCK) mode given that the image and adjacent channel-rejection requirements are reduced to their minimum. Additionally, due to the wideband nature of CCK channel, dc offset and $1/f$ noise can be uncomplicatedly removed by a highpass filter (HPF). However, it is not such straightforward in 802.11a/g orthogonal frequency-division multiplexing (OFDM) mode. An improper choice of pole frequency may result in a significant distortion of those close-to-zero subcarriers. Alternatively, increasing the IF to a value equals to half of the channel spacing (i.e., 10 MHz for a, 12.5 MHz for g) appears to be more effective since the image-rejection requirement at such a low-IF value (i.e., 30 dB) is still practical to achieve for an error vector magnitude (EVM) of –25 dB. In addition, a LIF can alleviate the trade-offs encountered in designing the highpass filter (HPF) for dc-offset cancellation. This idea is illustrated by comparing the ZIF (Figure 4-3(a)) and +LIF-to-BB (Figure 4-3(b)) downconversion of an OFDM channel. First, the cutoff frequency of the LIF-HPF_1 can be highly increased in compare with that of the ZIF-HPF, leading to significant area savings while shorting the settling time in dc-offset transients. Second, since the tolerable instability of the reference crystal is ±20 ppm, a mixed-mode automatic frequency control (AFC) is essential for ZIF [4.8]. The AFC estimates digitally the frequency error (which is as high as 214 kHz at 5.35 GHz) and then compensates it in the analog domain by offsetting the frequency of the RF local oscillator (LO_{RF}) by the same amount. In contrast, a LIF can endure much larger frequency errors at HPF_1 (i.e., 3.75 MHz). The compensation is therefore possible to be postponed to the IF LO (LO_{IF}), which has the simplicity of much lower operating frequencies (≤12.5 MHz).

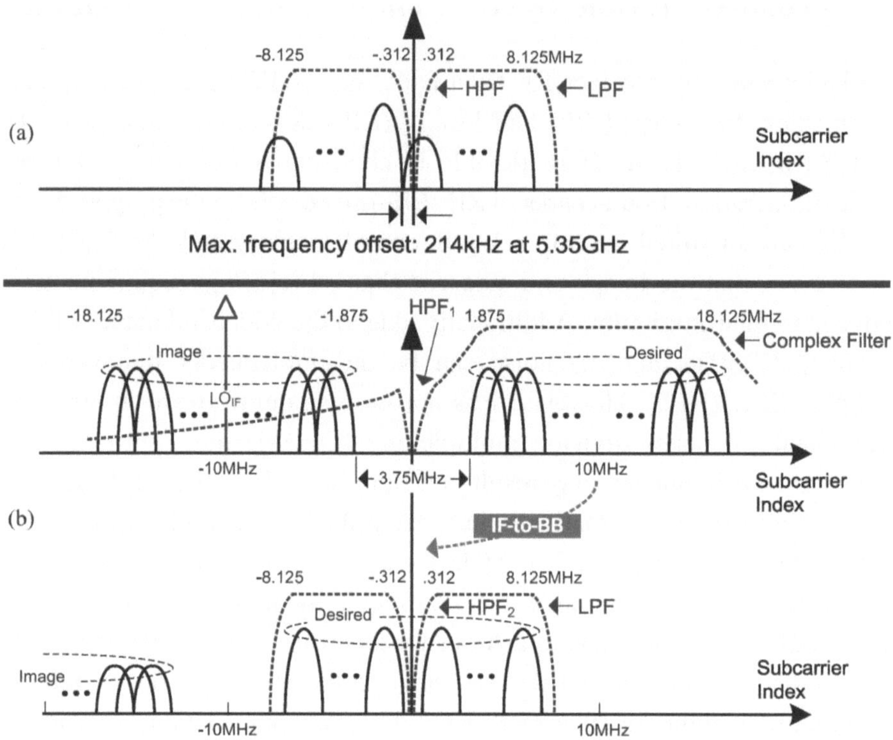

Figure 4-3. Downconversion of 5-GHz band OFDM channel: (a) ZIF and (b) +LIF-to-BB approaches

2.3 Proposed baseband-signal conditioning for cost-efficient reconfiguration

The efficiency of the aforesaid flexible-IF reception is critically determined by the permutation of functional blocks. The system partition and its operation in LIF mode are depicted in Figure 4-4(a) and (b), respectively. The analog BB interfaces the dual-band radio to the digital BB. The analog-BB signal conditioning starts with a center-frequency-tunable (i.e., at +IF, –IF or dc) complex filter (CF) (Chapter 3), followed by an IF-to-BB downconverter that is bypassable in ZIF mode, a channel-selection LPF and a PGA. In this way, the specifications of the LPF, PGA, and the A/D converter in the digital BB, are all idéntical in either mode, maximizing the functional-block sharing.

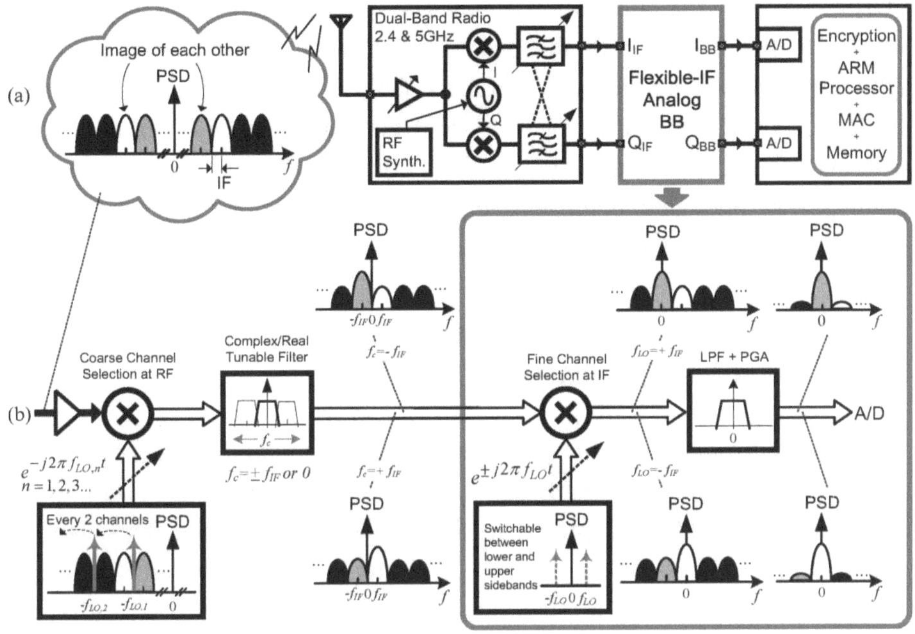

Figure 4-4. (a) System partition and (b) its operation in LIF mode

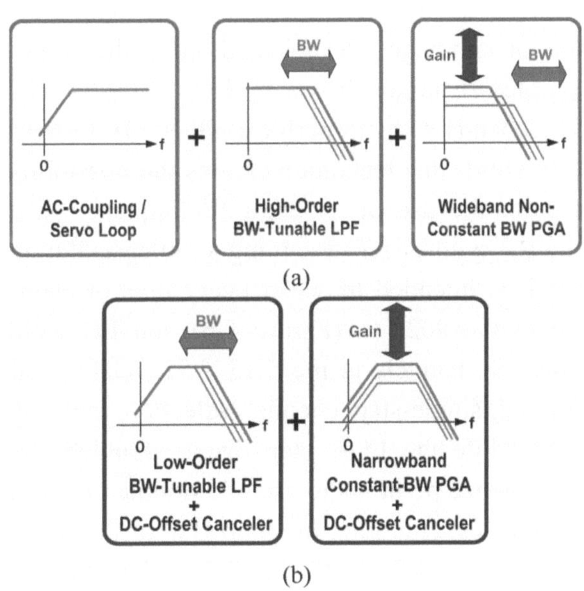

Figure 4-5. Gain-BW signal conditioning: (a) conventional and (b) proposed approaches

Different from the conventional gain-BW conditioning approach (Figure 4-5(a)), in this design both the backmost channel-selection lowpass filter (LPF) and programmable-gain amplifier (PGA) have individual highpass filter (i.e., HPF_2) embedded for local dc-offset cancellation (Figure 4-5(b)). By doing so, and realizing the PGA with a constant-BW gain adjustment, the system's BW and gain controls can be separately set while reducing the order of the LPF and the BW requirement of the PGA.

2.4 Proposed two-step channel-selection technique for radio front-end simplification

One may observe that the proposed receiver has dual frequency downconversions (i.e., RF→Flexible-IF→BB), permitting the use of *two-step channel selection* [4.9] (Chapter 3) to simplify the RF front-end. This claim is described in Figure 4-4(b) as well. In LIF mode, the selected IF (i.e., one-half the channel spacing) implies the desired channel and its first adjacent one are in image relationship. Owing to that, the RF-to-IF downconversion will translate both of them to the identical IF and conjugate their phases. With a tunable center frequency between ±IF, the CF can flexibly pass either the desired channel or its image. The selected one is then downconverted to dc by using a complex-IF mixer driven by LO_{IF}. LO_{IF} is made switchable between the two sidebands so as to perform either +IF-to-BB or –IF-to-BB downconversion. In short, this technique creates the possibility of selecting channel at IF without the use of an extra IF frequency synthesizer. The consequence to the RF synthesizer (assuming an integer-*N* type) is a relaxed specification since it is intended to select every *pair* of channel. Mapping this simplification to the 802.11a (Figure 4-6), ten LO locking positions suffice to cover the 19 channels in the 5.15–5.725 GHz band. Comparing with the conventional single-step channel selection, half of the locking positions are saved while the LO_{RF} step-size is doubled (from 20 to 40 MHz). A doubled step-size permits the use of a doubled reference frequency in the frequency synthesis, enlarging the loop BW of the phase-locked loop and reducing the division ratio in the modulus. The former results in a shorter settling time and lower LO_{RF} in-lock phase noise, whereas the latter reduces the LO_{RF} close-in phase noise. The trade-offs in designing frequency synthesizer have been extensively analyzed in [4.10].

Likewise, covering the three nonoverlapping channels in the 2.4-GHz band requires totally five synthesized carriers running with a 12.5-MHz step-size for ZIF and LIF downconversions.

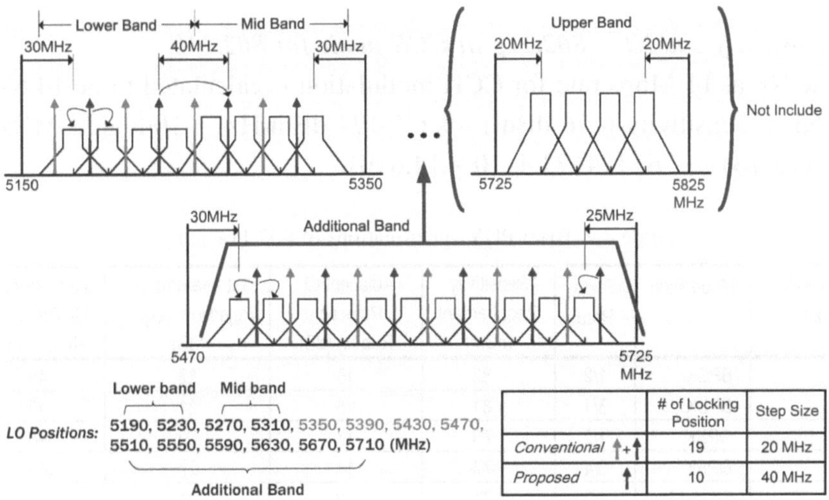

Figure 4-6. 5-GHz LO plan with and without using two-step channel selection

3. TRANSLATING THE 802.11A, B AND G STANDARDS TO RECEIVER DESIGN SPECIFICATION

This subsection summarizes mainly the detailed calculations of receiver (RX) noise figure (NF), input-referred second-order and third-order intercept points (IIP2 and IIP3). The information, as briefly summarized in Tables 4-1 and 4-2, are obtained from the physical layer (PHY) specifications of 802.11a, b and g [4.11][4.12][4.13]. In our case, only the CCK mode of b is considered as a part of g.

RX Sensitivity and NF – ***802.11a/g***

The required signal-to-noise ratios (SNRs) for 6-Mbps rate using binary phase shift keying (BPSK) modulation and 54-Mbps rate using 64-quadrature amplitude modulation (QAM64) are ~5 and ~20 dB, respectively. Ideally, the NF at 6 Mbps is calculated to be ~15 dB, i.e., NF=Sensitivity (–82 dBm) – kT (–174 dBm/Hz) – 10·log[16.3 MHz] – SNR (5 dB) = –82+174–72–5 = 15 dB. Likewise, the NF at 54 Mbps is calculated to be ~15.4 dB, i.e., NF =

Sensitivity (–65 dBm) – kT (–174 dBm/Hz) – 10·log[16.3 MHz] – SNR (21.6 dB) = –65+174–72–20 = 17 dB. Comparing the two results, the NF at 6-Mbps rate sets the design specification.

*RX Sensitivity and NF – **802.11g in CCK mode for 802.11b***

The NF at 11-Mbps rate for CCK modulation is calculated to be 14.6 dB, i.e., NF = Sensitivity (–76 dBm) – kT (–174 dBm/Hz) – 10·log[22 MHz] – SNR (10 dB) = –76+174–73.4–10 = 14.6 dB.

Table 4-1. Brief PHY specifications of 802.11a and g

Data Rate (Mbps)	Modulation	Code Rate	Sensitivity Requirement (dBm)	Adjacent Ch Rejection Requir. (dB)	1st Alternate Ch Rejection Requir. (dB)	2nd Alternate Ch Rejection Requir. (dB)
6	BPSK	1/2	–82	16	32	49
9	BPSK	3/4	–81	15	31	48
12	QPSK	1/2	–79	13	29	46
18	QPSK	3/4	–77	11	27	44
24	QAM-16	1/2	–74	8	24	41
36	QAM 16	3/4	–70	4	20	37
48	QAM-64	2/3	–66	0	16	33
54	QAM-64	3/4	–65	–1	15	32

Table 4-2. Brief PHY specifications of 802.11b

<table>
<tr><th>Data Rate (Mbps)</th><th>Modulation</th><th>Sensitivity Requir. (dBm)</th><th>Adjacent Ch Rejection Requir. (dB)</th></tr>
<tr><td>1</td><td>D-BPSK</td><td>–80</td><td rowspan="4">35</td></tr>
<tr><td>2</td><td>D-QPSK</td><td></td></tr>
<tr><td>5.5</td><td>CCK</td><td></td></tr>
<tr><td>11</td><td>CCK</td><td>–76</td></tr>
</table>

*RX Interferences, IIP3 and IIP2 – **802.11a***

The standards specify that the linearity test case should include one –66-dBm tone at 20 MHz and one –50-dBm tone at 40 MHz, while the desired signal is 3 dB above the sensitivity, i.e., –79 dBm. However, the above test case is not fully indicative in practice. For 802.11a, the second alternate channel can be considered as a –30-dBm blocker. Thus, assuming that there are two –33-dBm tones (resulting in a power sum of –30 dBm) at ~50 MHz and the speed is at 6-Mbps rate in an 802.11a network, then, the overall IIP3 at LNA input is calculated to be: –33+(|–82–5|–33)/2 = –6 dBm (802.11a).

For IIP2, it is calculated to be: –33+(|–82–5|–33)/1 = 21 dBm (802.11a). Similarly, assuming that there are two –33-dBm tones at ~50 MHz and the speed is at 54-Mbps rate. The overall IIP3 at LNA input is calculated to be: –33+(|–65–20|–33)/2 = –7 dBm. For IIP2, it is calculated to be: –33+(|–65–20| –33)/1 = 19 dBm. In overall, the IIP3 and IIP2 at 6 Mbps are the toughest requirements.

RX Interferences, IIP3 and IIP2 – ***802.11g in CCK mode for 802.11b***

The 802.11g in OFDM mode features the same linearity test case with 802.11a presented above. Differently in CCK mode, the test case should be executed with two –35-dBm tones (i.e., resulted in a power sum of –32 dBm) away from the desired channel by 25 MHz, while the desired channel is 6 dB above the sensitivity level, i.e., –70 dBm. However, when the linearity is encountered together with the SNR requirement, the overall IIP3 and IIP2 at LNA input should be calculated without the add-on 6 dB in desired channel, i.e., IIP3: –32+(|–76–10|–32)/2 = –5 dBm, and IIP2: –32+ (|–76–10|–32)/1=22 dBm.

Overall, the IIP3 and IIP2 in CCK mode and at 11-Mbps rate are the toughest. A summary of the toughest 802.11a/b/g standard requirements is listed in Table 4-3.

Table 4-3. 802.11a/b/g standard PHY requirements

Parameters	802.11b/g 2.4 GHz Band	802.11a 5 GHz Band
Noise Figure (dB)	14.6	15
IIP3 (in OFDM) (dBm)	–6	–6
IIP3 (in CCK) (dBm)	–5	
IIP2 (in OFDM) (dBm)	21	21
IIP2 (in CCK) (dBm)	22	

4. GAIN PLAN

The gain plan must leverage the NF and linearity (here, IIP3) with respect to each stage such that the signal swing arrived at the analog-to-digital (A/D) converter is optimized between 0 and –5 dBm. The proposed zero-IF/low-IF receiver plan is shown in Table 4-4. In the RF part, except the antenna (ANT), the selected values for the band-selection filter (BSF), dual-band low-noise amplifier (LNA) and dual-band RF mixer (MIX) are obtained from [4.14]. The other values planned for the CF, IF MIX, channel-selection lowpass filter (LPF) and programmable-gain amplifier (PGA) are obtained through the analysis presented in the next subsection.

Table 4-4. Receiver gain and linearity plan for 802.11a/b/g

Block	ANT	BSF	LNA	RF MIX	CF + IF MIX + LPF + PGA	Total	Specifications
Gain [dB]	–	–3	0...15	12	0...50	9...74	10...74
NF [dB]	–	3	3	11	30	9.39	9.5 (with 5.5-dB margin for fading)
IIP3 [dBm]	–	–	–2	+15	+15	–3.14	–5

5. SPECIFICATION OF THE ANALOG BASEBAND

First, it would be necessary to determine the gain and filter-order requirements for 802.11a and g because b requires a much higher filter-order requirement, when comparing the values from Tables 4-3 and 4-4 obtained from their standard PHY specifications. As it will be shown later, realizing the excessive adjacent channel rejection (i.e., ~19 dB) for b in the digital domain is more efficient [4.15]. The gain charts concurrently take the lowest and highest signal levels (with the presence of the adjacent and alternate channels) into account are listed in Figure 4-7(a) and (b), respectively. The dotted lines represent the signal levels of the adjacent and alternate channels after filtering. The maximum allowable signal level of 802.11a/b/g is –30/ –10/–20 dBm, respectively. The minimum signal level is 3 dB above the sensitivity for a and g, but 6 dB above the sensitivity for b. The band-selection filter (BSF) is assumed to have –3-dB gain loss. The LNA and RF MIX together offer two gain steps, i.e., 15 and 27 dB. The rest programmable gains are realized in the analog baseband channel-selection lowpass filter (CSF) and programmable-gain amplifier (PGA). They together should exhibit a gain range of 8–50 dB, a filter order of five (Butterworth for good inband linearity) and a cutoff of ~8 MHz.

In CCK mode, the cutoff is 7.5 MHz and additional filtering in the digital domain is required to fulfill the adjacent channel-rejection requirement of 35 dB (Figure 4-8(a) and (b)). Due to a higher maximum signal power level of –10 dBm, the LNA and MIX together should offer 12–27 dB.

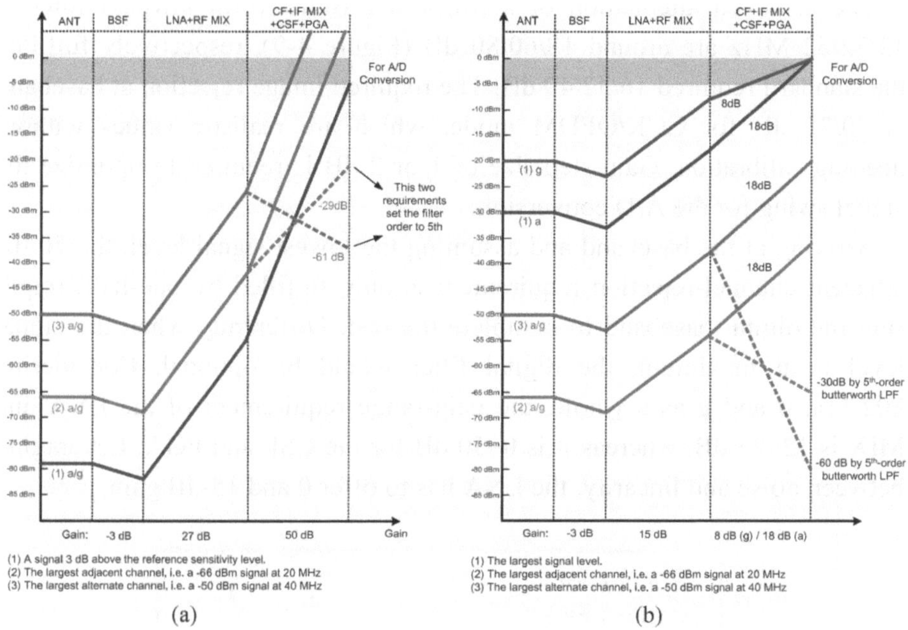

Figure 4-7. Gain and filter-order plan for 802.11a/g in OFDM mode at the (a) lowest and (b) highest gain levels.

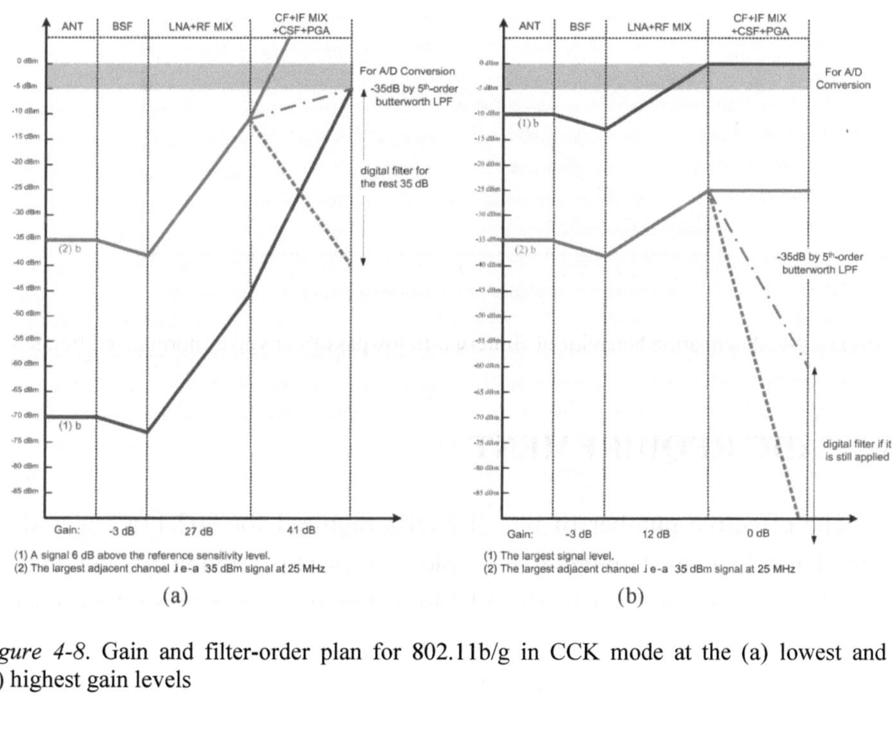

Figure 4-8. Gain and filter-order plan for 802.11b/g in CCK mode at the (a) lowest and (b) highest gain levels

The achieved attenuation of a fifth-order Butterworth lowpass filter at 12/32/52 MHz are around 18/60/80 dB (Figure 4-9), respectively fulfilled the standard required 16/32/49 dB. The required image rejection at baseband is 30/32 dB for CCK/OFDM mode, which are realistic values without needing calibration. Gain step-size of 1 or 2 dB is required to optimize the signal swing for the A/D conversion.

Arriving at the baseband and assuming the lowest signal level, the 70-dB adjacent channel-rejection requirement is only fulfilled by one-half, requiring the digital baseband to complete the rest. Differently when the signal level is at maximum, the digital filter would be optional. Considering 802.11a, b and g as a whole, the gain-range requirement of the LNA and MIX is 12–27 dB, whereas it is 0–50 dB for the CSF and PGA. Leveraging between noise and linearity, the LNA has to offer 0 and 15-dB gain.

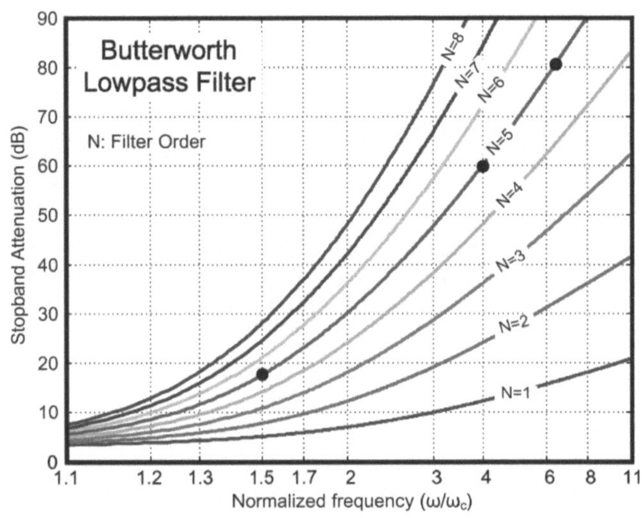

Figure 4-9. Attenuation behavior of Butterworth lowpass filter versus normalized frequency

6. ADC REQUIREMENT

The effective number of bits (ENOB) required for 802.11a/b/g is data-rate dependent, as tabulated in Table 4-5 [4.16]. Obviously, the highest ENOB requirement is set by the 54-Mbps data rate, suggesting the use of an ADC with minimally 9-bit resolution. Generally, a 10-bit 20-to-22-Ms/s ADC can fulfill the OFDM/CCK mode. However, if it is of pipelined topology, the latency in 10-stage is too long for preamble in OFDM mode. A sampling rate of 40-to-44 Ms/s befits more the applications; not mentioning it also helps relaxing the anti-aliasing filter. For state-of-the-art works, a 10-bit 40-MS/s pipelined ADC consumes 12 mW of power [4.17].

Table 4-5. ADC requirement

Standard	Data Rate	AWGN SNR with impairments	Quantization Bits	ENOB
802.11a/g	6 Mbps	4 dB	5	5
	9 Mbps	6 dB	5	5
	12 Mbps	7 dB	5	5
	18 Mbps	10.5 dB	5	5
	24 Mbps	11.8 dB	6	6
	36 Mbps	15.8 dB	6	6
	48 Mbps	22 dB	8	8
	54 Mbps	23 dB	8	8
802.11b	1 Mbps	4 dB	4	4
	2 Mbps	6.5 dB	4	4
	5.5 Mbps	10 dB	6	6
	11 Mbps	11.5 dB	6	6

7. SUMMARY

This chapter has described the system design of a SiP receiver for IEEE 802.11a/b/g WLAN. An optimum system partition in conjunction with a 3D floorplan has allowed an efficient SiP integration. The two-step channel selection technique has been practically considered and applied, allowing a relaxation of the RF synthesizer design specifications through a channel-selection partition between the RF front-end and analog BB. The overall receiver structure has been optimized for multistandard compliance by utilizing a ZIF reception for 802.11b/g in CCK mode and a LIF reception for 802.11a/g in OFDM mode. The derived design specifications are the basis of the realization of the analog BB and its circuitry, as it will be described next in Chapters 5 and 6.

REFERENCES

[4.1] S. Chou, "Integration and Innovation in the Nanoelectronics Era," *IEEE International Solid-State Circuits Conference (ISSCC), Digest of Technical Papers*, pp. 36–41, Feb. 2005.

[4.2] J. M. Rabaey, F. De Bernardinis, A. M. Niknejad, B. Nikolic and A. Sangiovanni-Vincentelli, "Embedding Mixed-Signal Design in Systems-on-Chip," in *Proc. IEEE*, vol. 94, pp. 1070–1088, June 2006.

[4.3] R. S. Patti, "Three-Dimensional Integrated Circuits and the Future of System-on-Chip Designs," in *Proc. IEEE*, vol. 94, pp. 1214–1224, June 2006.

[4.4] Nicky C. C. Lu, "Emerging Technology and Business Solutions for System Chips," *IEEE International Solid-State Circuits Conference (ISSCC), Digest of Technical Papers*, pp. 25–31, Feb. 2004.

[4.5] P.-I. Mak, S.-P. U and R. P. Martins, "Design and Test Strategy Underlying a Low-Voltage Analog-Baseband IC for 802.11a/b/g WLAN SiP Receivers," in *Proc. IEEE International Symposium on Circuits and Systems (ISCAS)*, pp. 2473–2476, May 2006.

[4.6] M. Bhardwaj, et al., "A 180MS/s 162Mb/s Wideband Three-Channel Baseband and MAC Processor for 802.11a/b/g," *IEEE International Solid-State Circuits Conference (ISSCC), Digest of Technical Papers*, pp. 454–455, Feb. 2005.

[4.7] M. Brandolini, P. Rossi, D. Manstretta and F. Svelto, "Toward Multistandard Mobile Terminals–Fully Integrated Receivers Requirements and Architectures," *IEEE Transactions on Microwave Theory and Techniques*, vol. 53, No. 3, pp. 1026–1038, Mar. 2005.

[4.8] A. Behzad, Z. M. Shi, S. Anand, L. Lin, K. Carter, M. Kappes, T.-H. Lin, T. Nguyen, D. Yuan, S. Wu, Y. C. Wong, V. Fong and A. Rofougaran, "A 5-GHz Direct-Conversion CMOS Transceiver Utilizing Automatic Frequency Control for the IEEE 802.11a Wireless LAN Standard," *IEEE Journal of Solid-State Circuits (JSSC)*, vol. 38, pp. 2209–2220, Dec. 2003.

[4.9] P.-I. Mak, S.-P. U and R. P. Martins, "Two-Step Channel Selection – A Novel Technique for Reconfigurable Multistandard Transceiver Front-Ends," *IEEE Transactions on Circuits and Systems-I (TCAS-I)*, vol. 52, No. 7, pp. 1302–1315, July 2005.

[4.10] B. Razavi, *RF Microelectronics*. Prentice-Hall, 1998.

[4.11] *Wireless LAN Medium Access Control (MAC) and Physical Layer (PHY) Specifications*—High-Speed Physical Layer in the 5 GHz Band, ANSI/IEEE Standard 802.11a, 1999.

[4.12] *Wireless LAN Medium Access Control (MAC) and Physical Layer (PHY) Specifications*—Higher-Speed Physical Layer Extension in the 2.4 GHz Band, ANSI/IEEE Standard 802.11b, 1999.

[4.13] *Wireless LAN Medium Access Control (MAC) and Physical Layer (PHY) Specifications*—Further Higher-Speed Physical Layer Extension in the 2.4 GHz Band, IEEE Standard 802.11g/D1.1, 2002.

[4.14] A. Behzad, "Wireless-LAN Radio Design," *IEEE International Solid-State Circuits Conference (ISSCC)*, Tutorial T2, Feb. 2004.

[4.15] P. Zhang, L. Der, D. Guo, I. Sever, T. Bourdi, C. Lam, A. Zolfaghari, J. Chen, D. Gambetta, B. Cheng, S. Gower, S. Hart, L. Huynh, T. Nguyen and B. Razavi "A CMOS Direct-Conversion Transceiver for IEEE 802.11a/b/g WLANs," in *Proc. IEEE Custom Integrated-Circuit Conference (CICC)*, pp. 409–412, Oct. 2004.

[4.16] D. Shoemaker, "Wireless LAN: Architecture and Design," *IEEE International Solid-State Circuits Conference (ISSCC)*, Tutorial, Feb. 2003.

[4.17] J. Arias, V. Boccuzzi, L. Quintanilla, L. Enríquez, D. Bisbal, M. Banu and J. Barbolla, "Low-Power Pipeline ADC for Wireless LANs," *IEEE Journal of Solid-State Circuits (JSSC)*, vol. 39, pp. 1338–1340, Aug. 2004.

Chapter 5

LOW-VOLTAGE ANALOG-BASEBAND TECHNIQUES

1. INTRODUCTION

The rapid downscaling of CMOS technology has accelerated the integration of complete wireless systems on a single chip. The associated reliability and leakage-current issues, regrettably, have driven the downsizing of power supply much faster than that of transistor's threshold voltage, continuously shrinking the voltage headroom for analog-circuits design. In view of that, techniques to attain a low-voltage (LV) operation have been extensively investigated, covering from elementary transistors for basic building blocks, to circuit structures for specific analog functions. For instance, a LV operational transconductance amplifier, or generally an operational amplifier (OpAmp), can be realized with multi-threshold [5.1], floating-gate [5.2], bulk-input [5.3] or bulk-biased [5.4] transistors; whereas a LV switched-capacitor (SC) delta-sigma modulator can be built with clock-boosting/switched-OpAmp [5.5], reset-OpAmp [5.6] and switched-resistor-capacitor (*RC*) [5.7] techniques.

This chapter deals with the design of low-voltage continuous-time (CT) analog-baseband circuitry. Basic circuit elements: OpAmp, CT level shifter, linear resistor-to-current (*R*-to-*I*) converter, common-mode feedback circuit (CMFB), current switch and MOS capacitor are described first. Novel analog functions, namely inside-OpAmp dc-offset canceler (DOC), double-quadrature downconverter based on a series-switching (SS) mixer-quad and a multi-phase *I*/*Q* generator, channel-selection filter (CSF), and switched current-resistor (SCR) programmable-gain amplifier (PGA), are then proposed. The reinforced techniques demonstrate that down to a 1-V supply, attaining high performance do not always implies an increase in power consumption or manufacturing cost for specialized devices.

2. OPERATIONAL AMPLIFIER (OPAMP)

Under low-voltage constraints, cascode structures are no longer adequate for achieving a generally sufficient dc gain of ~60 dB because of a very small signal swing will result. While cascade structures not only meet the dc gain, but also maintain a large signal swing for a maximum signal-to-noise ratio, at the expense of a lower gain-bandwidth product. Figure 5-1 shows a typical folded-cascode p-type 2-stage Miller-compensated OpAmp. The maximum common-mode (CM) voltage at V_{inp} (V_{inn}) is $V_{DD}-2V_{SDsat}-|V_{T,p}|$, where V_{SDsat} is one saturation voltage. For instance, with a CM level of 0.1 V at V_{inn} (V_{inp}), $|V_{T,p}|$ of 0.7 V and V_{SDsat} of 0.1 V, the minimum V_{DD} required is 1 V. It is noteworthy that the I/O common-mode levels are unmatched since the output CM level should be midway the V_{DD} to maximize the output swing (i.e., $V_{DD}-V_{SDsat}-V_{DSsat}$). Adding a level shifter to V_{inn} (V_{inp}) allows the input CM level to be biased differently from the output CM, as described next.

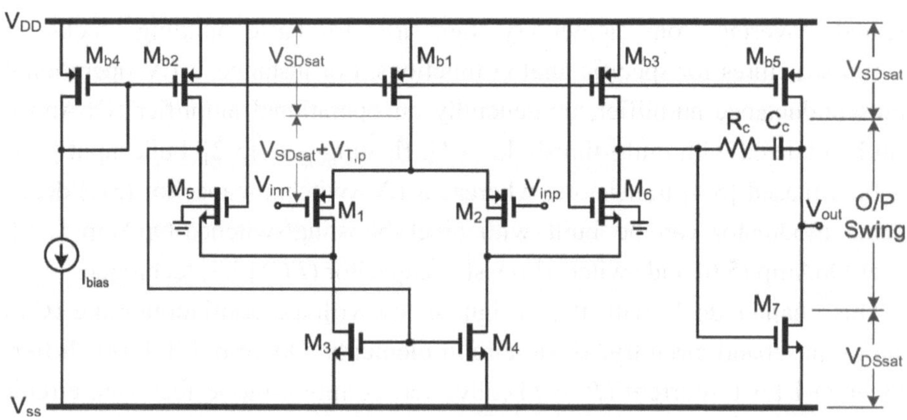

Figure 5-1. A typical folded-cascode p-type 2-stage Miller-compensated OpAmp

3. CT LEVEL SHIFTER

Unlike the SC level shifter [5.6] that is especially effective for discrete-time OpAmp-based circuits, a CT level shifter, on the other hand, can be as simple as a resistor (Figure 5-2(a)) or a current source (Figure 5-2(b)). With the I/P and O/P CM levels biased at $V_{DD}/2$, the CM level at v_x is defined by,

$$v_x = \frac{R_b \frac{V_{DD}}{2}}{R_{fb} // R_{ff} + R_b}, \tag{5.1}$$

where R_{ff} and R_{fb} are the forward and feedback resistors, respectively. R_b is the resistor for level shifting in Figure 5-2(a) or represents the output resistance of the current sink I_b in Figure 5-2(b). It is noteworthy that both types of level shifter for p-type OpAmps. For n-type ones, R_b in Figure 5-2(a) should not be connected to the ground, but V_{DD}, whereas I_b in Figure 5-2(b) should not be a current sink, but current source from the V_{DD}.

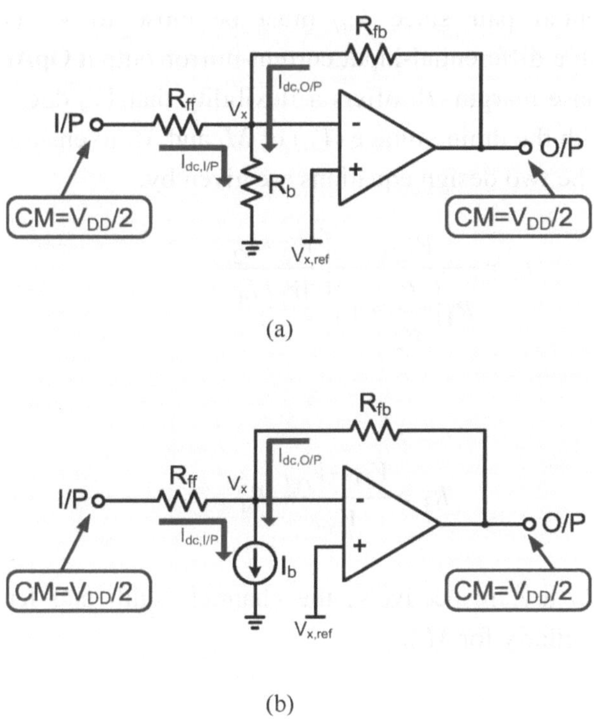

Figure 5-2. LV inverting amplifier for p-type OpAmps using (a) resistor and (b) current source as level shifters

The limitations of using a resistor as level shifter are two. First, R_b, R_{ff}, and R_{fb}, should be precisely matched to well define the CM at V_x. Second, the feedback factor, $R_{ff}/(R_{ff}+R_{fb}+R_b)$, is reduced, resulting in smaller bandwidth and longer settling time. To alleviate the latter problem, a current source can be employed instead because of its much higher output resistance. A current source, regrettably, requires a V_{DSsat} of minimally 0.1 V, brings on $1/f$ noise on top of the thermal noise, while demanding a resistor-to-current (R-to-I) converter to ensure the biased voltage enjoys process, voltage and temperature (PVT) immunity.

4. LINEAR R-TO-I CONVERTER

Depicted in Figure 5-3(a) and (b) are two types of *R*-to-*I* converters that can be used for current source and sink, respectively. Resistors R_{1-4} should be of the same type (e.g., polysilicon) to ensure a precision matching. The error amplifier A_{error} for the former (latter) requires an input stage of PMOS (NMOS) differential pair since V_{ref} must be close to V_{ss} (V_{DD}). A_{error} implemented with a differential-input current-mirror-output OpAmp exhibits an acceptable phase margin. R_3 offers a flexibility that V_{ref} does not require to be identical with the drain voltage (V_D) of M_1 and M_2 to ensure a precisely copied current. The two design equations are given by,

$$I_b = \frac{V_{DD}}{R_4\left(\frac{R_1}{R_2}+1\right)}\left(\frac{W_2/L_2}{W_1/L_1}\right), \tag{5.2}$$

and

$$R_3 = \frac{V_D - V_{ref}}{V_{ref}} R_4 , \tag{5.3}$$

where W_1 and L_1 are, respectively, the channel width and length of the transistors M_1 (similarly for M_2).

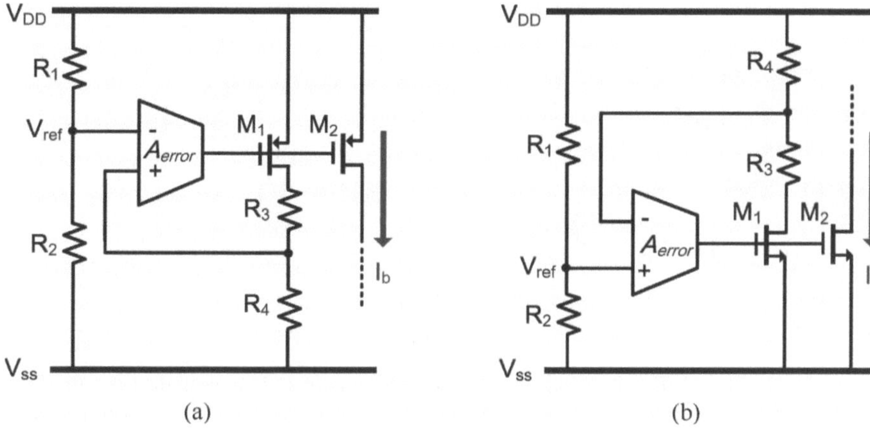

Figure 5-3. LV R/I converters for (a) current source and (b) current sink as the outputs

5. CT CMFB

For fully differential circuits, the level shifter can be substituted by an input common-mode feedback circuit (I-CMFB). As shown in Figure 5-4, assuming a p-type OpAmp, an error amplifier featuring a resistive input (or capacitive to exclude $R_{cm,in}$ in defining the feedback factor) biases VG+ and VG- by controlling the current source I_b in a feedback loop. The expense of an I-CMFB is worth since the dc-level of VG+ and VG- are flexible by controlling $V_{x,ref}$. The resistance value of $R_{cm,in}$ is critical in terms of feedback factor. The continuous-time feedback loop is insensitive to PVT variation. The current sources, M_{i7} and M_{i8}, behave like the 2nd stage of the OTA, thus the entire feedback loop has two dominant poles. The feedforward capacitor C_x is to improve the phase margin.

For the OpAmp's output, due to a large signal swing and since the common-mode level is $V_{DD}/2$, an output CMFB (O-CMFB) with resistive inputs for current sensing can be employed, as shown in Figure 5-5. Capacitors $C_{cm,out}$ may be required to improve the O-CMFB loop phase margin so that common-mode oscillations cannot be sustained. Eliminating the differential pair, the OTA has low input impedance. Depending on the speed requirements and the final feedback node, the stability is the main concern since the O-CMFB already experienced an extensive phase shift.

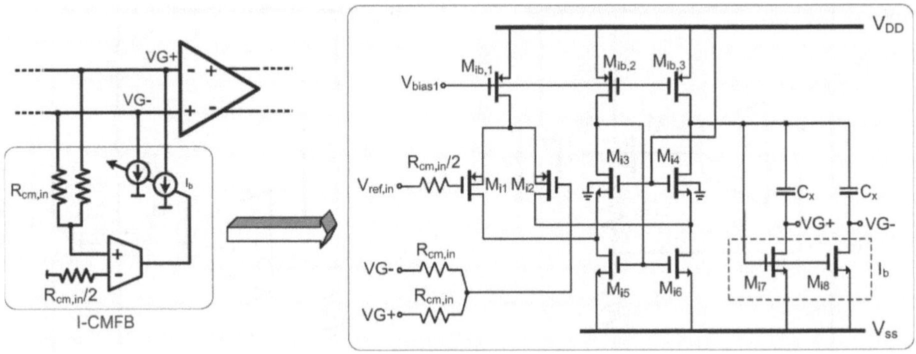

Figure 5-4. LV I-CMFB using a resistor detector along with a voltage-mode error amplifier

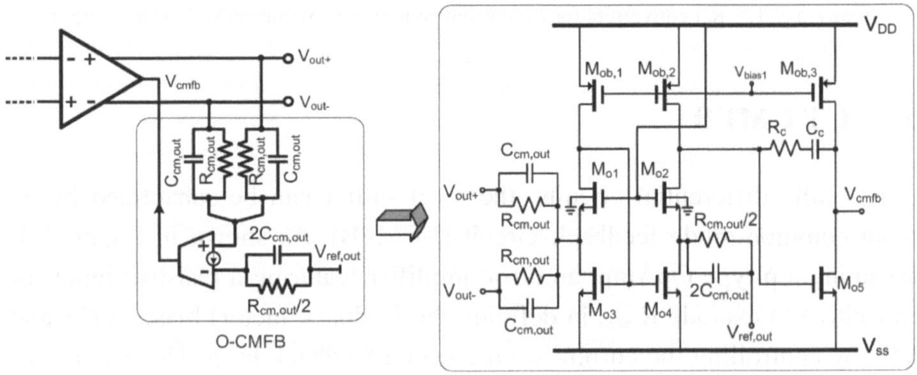

Figure 5-5. LV O-CMFB using a resistor detector along with a current-mode-input error amplifier

6. CURRENT SWITCH

The linearity of voltage switches is highly related with the overdrive voltage, V_{GS}–V_{th}. Current switch, on the other hand, features a good linearity under LV constraints. Again, assuming a p-type OpAmp, circuits like level shifter, multiplexer, mixer or track-and-hold amplifier [5.8], can be built by selecting appropriately the nodes of switching, as shown in Figure 5-6(a), (b) and (c), respectively.

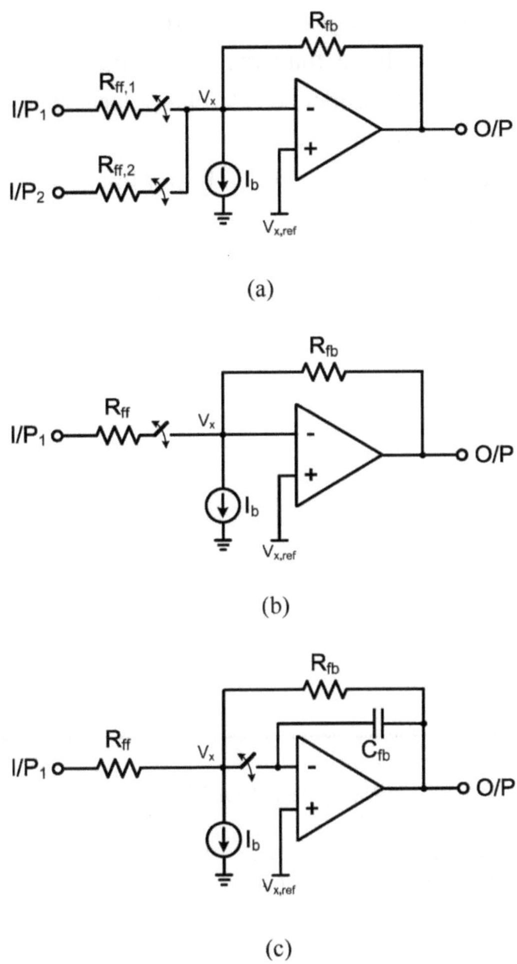

Figure 5-6. LV (a) multiplexer, (b) mixer and (c) track-and-hold amplifier

7. MOS CAPACITOR

In standard digital CMOS process, metal layers and MOSFETs are the primary choices for realizing capacitors. The former in parallel-plate or metal-wall configuration is highly linear but offering very low capacitance per unit area. The latter, in contrast, features a large capacitance per unit area (because their dielectric layer is constituted by a very thin gate oxide) at the expense of voltage-dependence charge distributions, i.e., accumulation, depletion and inversion (Figure 5-7). In most of the cases only the linear parts of the capacitance-voltage (CV) curve can be utilized, such as the inversion and accumulation regions. Operating in the inversion region requires a lower biasing voltage than the accumulation one, but inversion region shows a frequency-dependence CV characteristic at high frequency. Under a low-voltage supply of 1 V, both the inversion and accumulation regions cannot be used, leaving the relatively-nonlinear and low-capacitance depletion region for exploitation.

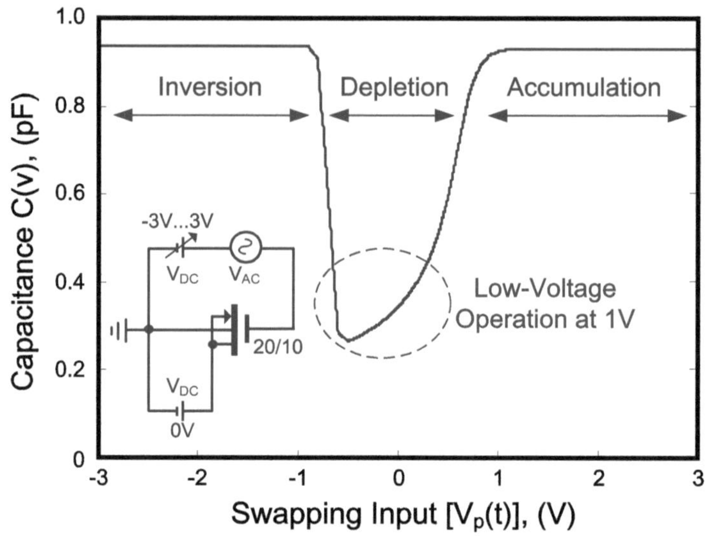

Figure 5-7. Typical quasi-static (low-frequency) CV characteristic of a p-channel MOS capacitor in a 0.35-μm CMOS process (t_{ox} = 7.6 nm)

7.1 Parallel-compensated depletion-mode MOS capacitor

Operating at the depletion region requires certain compensations, like series and parallel techniques [5.9], such that the CV characteristic can be linearized through substrate biasing, which leads to an extension of the linear voltage range due to the body effect. The former features a better linearity than the latter but lower capacitance per unit area. The effectiveness of either compensation is matching dependent, but not strongly process dependent. To be presented in latter sections, a parallel-compensation depletion-mode MOS capacitor is a good candidate for differential implementation of the proposed dc-offset canceler (DOC), given that the DOC is a *gm-C* integrator followed by a second gain stage. Thus, the required signal swing associated at the integration capacitor is very small, giving minimal nonlinearity penalty and permitting exploiting the capacitor in differential mode for capacitance doubling.

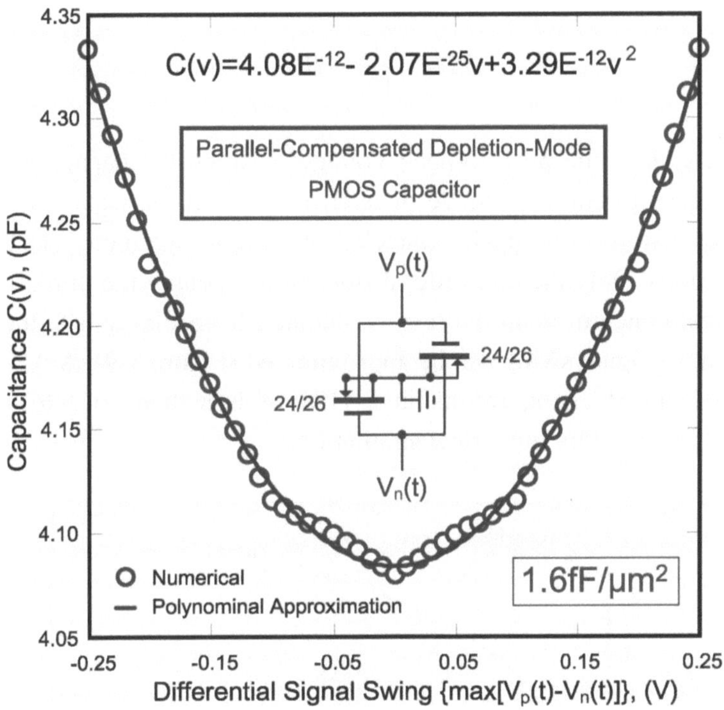

Figure 5-8. CV characteristic of a parallel-compensated p-channel MOS capacitor in depletion region

The simulated CV characteristic of the integration capacitor is shown in Figure 5-8. The CV dependency within a signal swing of 0.2-V_{pp} differential is 4.9%/V. The temperature capacitance dependency is 0.12%/°C. The capacitance per unit area in typical case is 1.6 fF/μm^2 (1.40 ~ 1.75 fF/μm^2 in process corners). It is equivalent to 3.2 fF/μm^2 in differential connection, and around 2× larger than that offered by the poly-poly capacitors (i.e., 0.86 fF/μm^2). Without using the complicated modeling equations as described in [5.9], the amount of nonlinearity can be estimated in a simpler way by applying a second-order polynomial approximation to the CV curve (i.e., $C(v) = C_0 + C_1 v + C_2 v^2$). The resulting coefficients determine the second-order (HD_2) and third-order (HD_3) harmonic distortions as given by [5.10],

$$HD_2 \approx \frac{C_1}{4C_0} B \ , \tag{5.4}$$

and

$$HD_3 \approx \frac{C_2}{12C_0} B^2, \tag{5.5}$$

respectively. *B* is the peak-to-peak voltage between the capacitor's inputs. Since there is no common-mode difference between the capacitor's inputs, even-order distortion is ideally canceled out due to symmetry (i.e., parallel compensation). Only the odd-order distortion (especially the third harmonic) is of critical concern. With the results obtained from Figure 5-8, the HD_3 to the amount of signal swing can be approximated (Figure 5-9). A 0.2-/0.1-V_{pp} differential signal swing results in a HD_3 of less than –52/–67 dB. The accuracy is verified by numerical simulation.

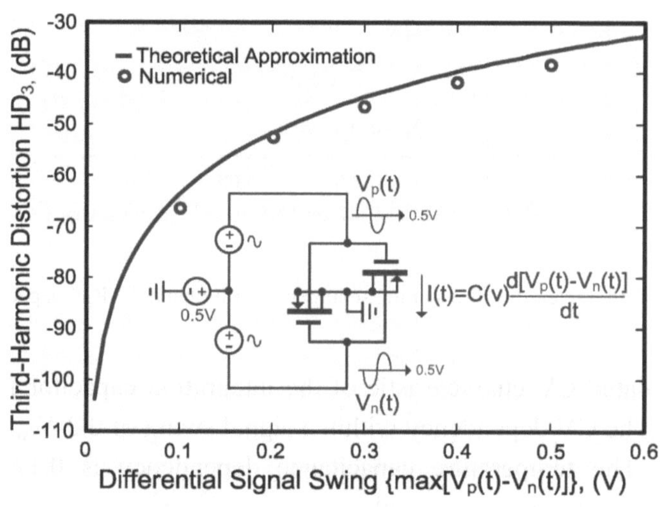

Figure 5-9. HD3 of a parallel-compensated p-channel MOS capacitor in a 0.35-μm CMOS process

Although parallel-compensated depletion-mode MOS capacitors benefit from differential connection and can cancel out all even-order harmonics, the common-mode voltage should be cautiously selected such that the signal never leads to forward bias in the transistor p-n junctions. As shown in Figure 5-10, the minimum voltage must be larger than, with a safe margin, the ground level minus the diode voltage V_D. Nevertheless, if the capacitor is employed at the output of an amplifier for integration, the common-mode voltage is mostly biased at one-half of the supply (e.g., 0.5 V at a 1-V supply) for maximizing the signal swing; the problem of forward bias, in this case, is avoided.

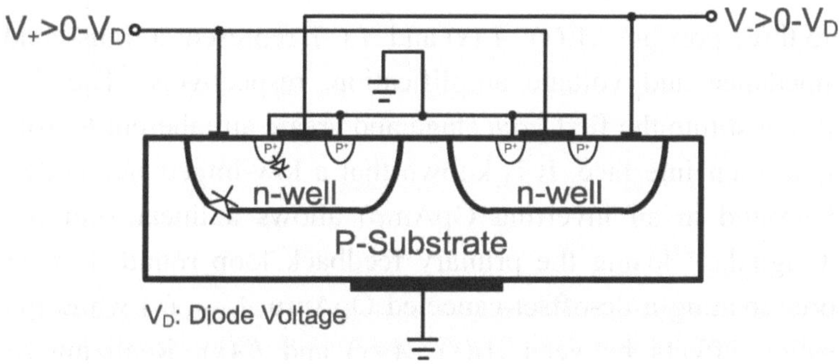

Figure 5-10. Cross section of a parallel-compensated p-channel MOS capacitor

8. INSIDE-OPAMP DC-OFFSET CANCELER (DOC)

DC-offset cancellation is a costly and complex issue for high-gain circuitry. A highpass pole with a large time constant prevents the baseband signal from intersymbol interference (ISI), but imposes a large chip-area impact and degrades the receiver's settling time. This section describes an innovative inside-OpAmp approach [5.11][5.12]. Realized using certain LV circuit techniques, a *gm-C* integrator-based DOC is embedded in an OpAmp to sense its differential output. The integrated imbalance is then amplified, and negatively fed back to the OpAmp at an inherent low-impedance node, leading to a switchable, compact, low-noise, linear and fast-convergent-speed dc-offset cancellation. The beneficial features, design details and circuit implementations are summarized as follows.

8.1 Basic principle 1 – design for switchability

Switchable DOC is commonly utilized to reduce the inconstant dc-offset-induced transient time and glitch noise by appropriately switching the high-pass pole(s) to a lower/higher frequency (e.g., switched transconductors

[5.13] and successive switching [5.14]). The quality of the operation is determined by the switchability of the DOC itself, and the disturbance to the original midband and high-frequency behaviors of the forward-path circuitry.

The switchability and disturbance of the proposed inside-OpAmp DOC are described in Figure 5-11 by using the inherent signal-conversion (i.e., voltage (V) and current (I)) characteristic of a 2-stage OpAmp $A_{OL}(s)$. Divided into three portions, $A_1(s)$, $A_2(s)$ and $A_3(s)$, represent a trans-conductance, transimpedance and voltage amplification, respectively. The former two portions constitute the first *gain* stage and create an inherent low-impedance node x_L at their interface. It is known that a low-impedance node (e.g., the virtual ground in an inverting OpAmp) allows a linear sum of multiple current signals. Closing the primary feedback loop round $A_2(s)$ and $A_3(s)$, therefore, creating a dc-offset-canceled OpAmp $A_{OL,OC}(s)$ while minimizing the loading effects between $A_1(s)$, $A_2(s)$ and $\beta_1(s)$. Realizing $\beta_1(s)$ as a transconductance integrator directly complies with the OpAmp's internal signal conversion and create a unilateral low-frequency feedback path from x_O to x_L.

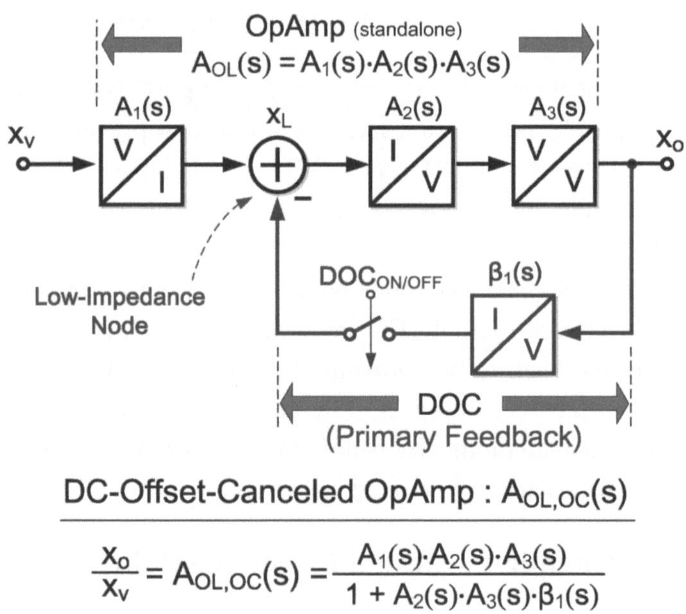

Figure 5-11. Internal signal conversion of a 2-stage OpAmp with DOC feedback

8.2 Basic principle 2 – negative feedback for noise and nonlinearity reduction

In addition to the switchability concern, applying the DOC feedback at node x_L rather than the commonly employed virtual ground can lower the noise and nonlinearity induced by the DOC. This is illustrated more generally in Figure 5-12 using an inverting amplifier. $\beta_1(s)$ resides on the forward path closed by the feedback resistor R_{fb} that creates another loop gain. As a result, the input-referred noise of $\beta_1(s)$ is divided by the preceded $A_1(s)$, which is a wideband transconductance amplifier. Likewise, the nonlinearity of $\beta_1(s)$ is suppressed by the OpAmp-R_{fb}-created loop gain, which is frequency independent and normally expressed as a feedback factor β_{IA}.

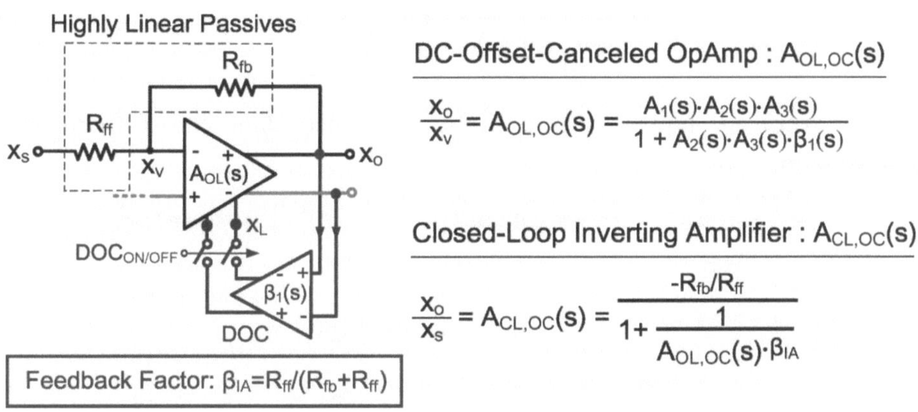

Figure 5-12. Inverting amplifier using OpAmp with built-in DOC

8.3 Basic principle 3 – negative feedback for area savings

This principle is described by Figure 5-13, the constitution of an inverting amplifier in frequency domain. The basic property of negative feedback in BW-extension [5.15] is that any highpass (lowpass) pole is shifted to a lower (higher) frequency value by an amount of the loop gain, (i.e., f_{LP}–$f_{LP,fb}$ for lowpass and f_{HP}–$f_{HP,fb}$ for highpass). It implies that the dc-offset-canceled OpAmp $A_{OL,OC}(s)$ used in closed loop (i.e., $A_{OL,OC}(s)$) can lower its highpass pole with no area overhead. It is differed from the traditional single DOC round multiple closed-loop stages, where the lower cutoff needs to be frequently adjusted once the forward path changes gain. The current approach, instead, rounds just one single stage, and the lower cutoff depends simply on the feedback factor of the closed-loop circuit. As it will be shown in latter subsections during the design of a PGA, the feedback factor is capable to be stabilized against gain, leading to a stable cutoff. The obtainable rejection at dc is given by,

$$\frac{\left|A_{CL,OC}(f_{HP,fb})\right|}{\left|A_{CL,OC}(f_z)\right|} \approx \left|\frac{\beta_1(f_z)}{A_1(f_z)}\right| \frac{1}{\beta_{IA}} . \tag{5.6}$$

One may wonder why $\beta_1(s)$ was not fed back at the virtual ground because it will give a higher rejection at dc, by a gain factor of $|A_1(f_z)|$, as given by,

$$\frac{\left|A_{CL,OC}(f_{HP,fb})\right|}{\left|A_{CL,OC}(f_z)\right|} \approx \frac{\left|\beta_1(f_z)\right|}{\beta_{IA}} . \tag{5.7}$$

The overhead, however, is a higher and inconstant cutoff frequency against gain.

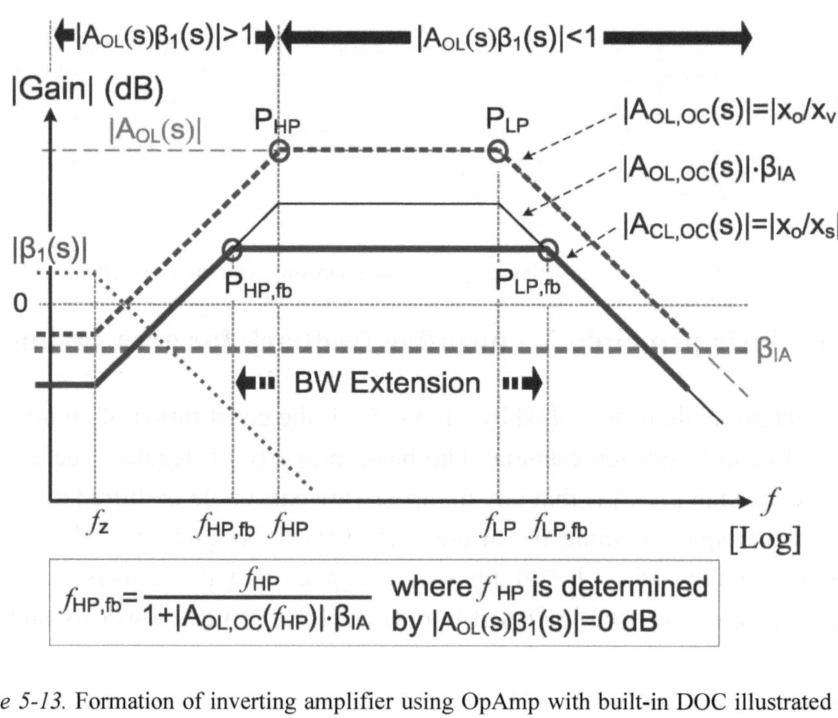

Figure 5-13. Formation of inverting amplifier using OpAmp with built-in DOC illustrated in frequency domain

8.4 Block-level design – convergent speed, stability and coverable range

Figure 5-14(a) shows the block schematic of the proposed OpAmp with a built-in DOC. $A_1(s)$, $A_2(s)$ and $A_3(s)$ are the corresponding sub-amplifiers of the OpAmp, when referring to Figure 5-11 and $\beta_1(s)$ is the DOC. The front-end resistors R_{oc}, sense the high swing outputs, V_{outp} and V_{outn}, with two differential-input single-ended-output current amplifiers $A_i(s)s$. The two $A_i(s)s$ drive the capacitor C_{oc} differentially forming a pseudo-differential *gm-C* integrator and obviating systematic dc offset, while offering common-mode rejection internally. Their low-impedance inputs allow $R_{oc}s$ to be cross-coupled between the two $A_i(s)s$ for better matching.

The output stage is a source-/sink-exchangeable charge pump I_{oc+} (I_{oc-}) for doubling the speed in canceling the dynamic dc offset (Figure 5-14(b)) and extending the output swing to rail-to-rail. It is worth mentioning that under a low V_{DD} (i.e., $V_{DD}<V_{T,n}+|V_{T,p}|$), there exists a dead zone (i.e., $V_{DD}-|V_{T,p}| < V_{oc+}<V_{T,n}$) where I_{oc+} and I_{oc} are cut off, implying that there would be no feedback in this region. In terms of stability, the dead zone prevents I_{oc+} (I_{oc-}) wandering between source and sink modes when V_{os} is extremely small. When there is no dc offset being determined, the common-mode voltage of V_{oc+} (V_{oc-}) will be automatically stabilized inside $V_{DD}-|V_{T,p}|<V_{oc+}<V_{T,n}$ to cut the feedback due to a large loop gain (i.e., $A_2(s)A_3(s)\beta_1(s)$), freeing V_{oc+} (V_{oc-}) from CMFB. The consequences of no CMFB, however, are two: (1) I_{oc+} (I_{oc-}) will fluctuate in canceling the dc offset, drawing an inconstant common-mode current from the OpAmp; (2) The common-mode rejection ratio (CMRR) cannot be further boosted on top of the *gm-C* integrator.

Alternatively, in terms of irremovable systematic dc offset, the size of the dead zone after input-referring to the OpAmp's output is given by (assuming a purely on/off operation at the threshold),

$$\frac{V_{DD}-|V_{T,p}|}{A_v} < V_{os} < \frac{V_{T,n}}{A_v}\,, \tag{5.8}$$

where A_v is the dc gain experienced from V_{outp} to V_{oc+}. A_v, is a high gain value to implement $\beta_1(s)$ as a large time-constant integrator, implying that the dead zone would be very small.

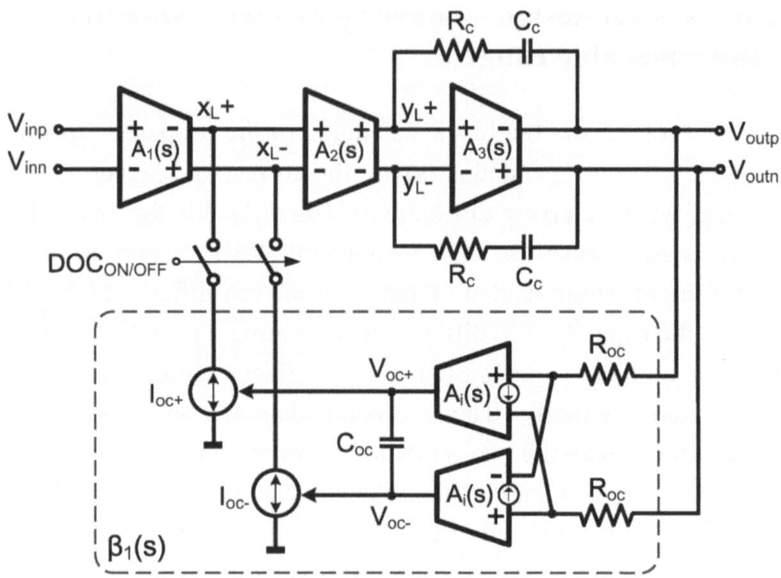

(a)

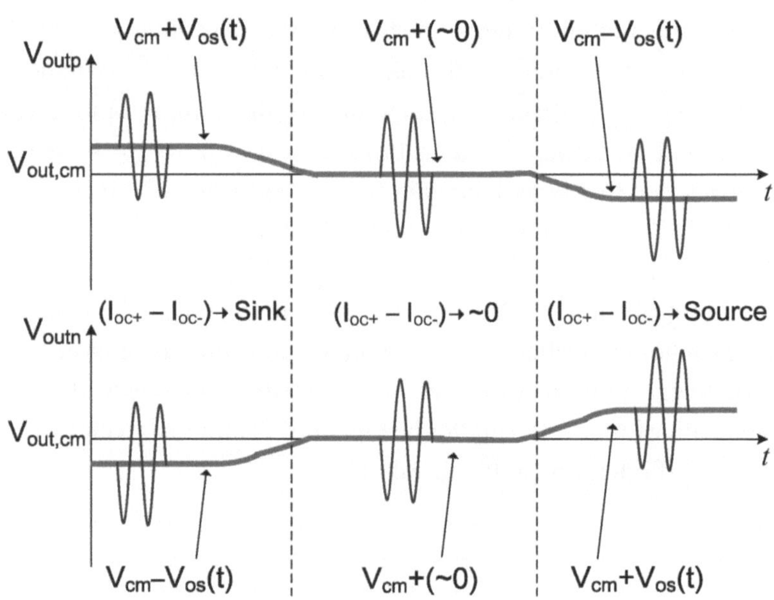

(b)

Figure 5-14. (a) Block schematic of OpAmp with built-in DOC and (b) its operation in canceling dynamic dc offset

8.5 Transistor-level implementation

Figures 5-15(a) and (b) show the schematics of the OpAmp and DOC, respectively. $A_1(s)$ is a p-channel differential pair. The gate of M_1/M_2 is biased at one V_{DSsat} (0.1 V), implying the minimum supply voltage (V_{DD}) is around 1 V (i.e., $V_{DD} \geq |V_{T,p}| + 2V_{SDsat} + V_{DSsat}$). A cross-coupled active load (M_{3A}/M_{3B} and M_{4A}/M_{4B}) [5.16] forms a wideband n-channel folded-cascode intermediate stage. $A_2(s)$ is a common-gate amplifier (M_5/M_6). Its low input impedance (at x_L+) offers a good current summation node for the DOC. On the other hand, its output impedance at y_L+ is only high for differential signal, eliminating the need of common-mode control. Only the final-stage common-source amplifier $A_3(s)$ is involved in the output CMFB, resulting in better stability. The phase margin is optimized by adding C_{cp} in $A_2(s)$ and adding R_c and C_c (i.e., Miller compensation) in $A_3(s)$.

To realize a large time constant, in the order of 0.1 ms, two circuit techniques were applied to the DOC. The first one is the use of self-biased subthreshold cascode current mirror for realizing the $A_i(s)$. As shown in the left-hand side of Figure 5-15(b), M_{oc5} and M_{oc6} are biased in the saturation region to absorb the dc current (~10 μA) from V_{outp} and V_{outn}, respectively. Due to the body effect associated with M_{oc9} and M_{oc10}, they are easy to be biased into the subthreshold region by using long channel-length devices for M_{oc13} and M_{oc14}. It is known that a subthreshold-biased MOS transistor offers a very high intrinsic dc gain that is independent of device geometry [5.17], making it highly appropriate for realizing a large time-constant integrator on-chip.

The second technique is a sink-/source-exchangeable charge pump (M_{oc1} and M_{oc2}). With it served as the output stage, not only the linearity requirement of $A_i(s)$ is relaxed, but also the signal swing associated at C_{oc}. Low signal swing enables C_{oc} to be implemented by weakly-nonlinear depletion-mode MOS capacitors (M_{oc17} and M_{oc18}) for area savings. The linearity consideration of MOS capacitor under low-voltage constraints has been given in this chapter *Section 7.1*. Operating inside the secondary feedback and using a reversed polarity in the connection [5.9], the nonlinearity penalty of using MOS capacitors will be minimized.

Breaking the feedback loop of $\beta_1(s)$, the s-domain transfer function of the DOC standalone is given by,

$$\frac{I_{oc+}(s)-I_{oc-}(s)}{V_{outp}(s)-V_{outn}(s)}=2\frac{gm_{oc}}{R_{oc}}\frac{A_{i,dc}r_{o,Ai}}{(sr_{o,Ai}C_{oc}+1)}, \tag{5.9}$$

where $A_{i,dc}$ and $r_{o,Ai}$ are the current-to-current dc gain and output resistance of $A_i(s)$, respectively. gm_{oc} is the transconductance of I_{oc+} (either M_{oc1} or M_{oc2}). The previously mentioned A_v is also given by (5.9) with gm_{oc}=1. Controlling the C_{oc} can minimize the corner frequency without disturbing the gain while the rest of the parameters have to be designed in parallel. The front-end R_{oc} dominates the DOC-induced noise.

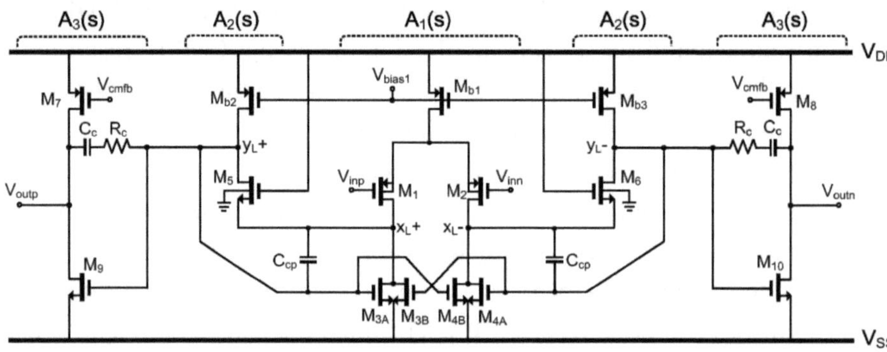

(a)

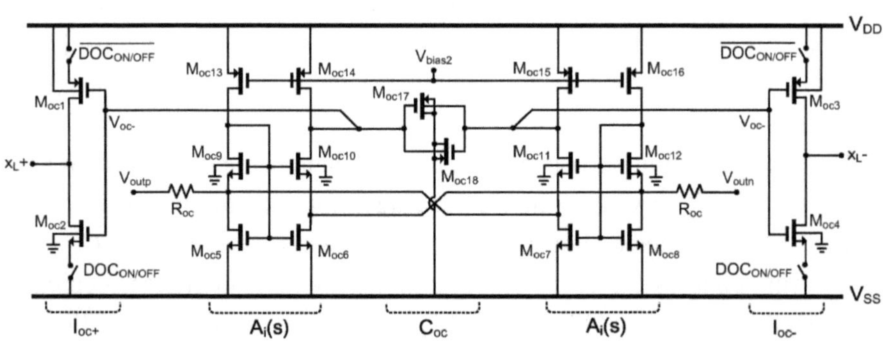

(b)

Figure 5-15. Full-circuit schematics: (a) OpAmp and (b) DOC (symbols correspond to Figure 5.14(a))

Conventionally, component mismatch inside the DOC *directly* limits the dc-offset cancellation. Differently here, the intrinsic dc offset of the DOC after input referred to the OpAmp's input is lowered by $A_I(s)$. The residual becomes part of the OpAmp's dc offset that will be multiplied by $1+R_{fb}/R_{ff}$ at the PGA's output. With $A_I(s)$ (i.e., a differential pair) offering a dc gain of around 25 dB, the dc offset induced by the DOC is of minor level when compared with that of the OpAmp.

Nevertheless, the DOC is designed to be differentially fully symmetric. The current branch of (M_{oc5}, M_{oc9} and M_{oc13}) has to be matched precisely with that of (M_{oc8}, M_{oc12} and M_{oc16}), and similarly, it must happen also with (M_{oc6}, M_{oc10} and M_{oc14}) and (M_{oc7}, M_{oc11} and M_{oc15}). Moreover, a large gain requirement of $A_i(s)$ requires M_{oc9-12} to be sized with a large aspect ratio, and $M_{oc13-16}$ to be sized with long channel length.

8.6 Simulation results

The sizes of the components employed in the OpAmp and DOC loop are summarized in Table 5-1. The open-loop AC responses of the OpAmp with DOC simulated in the typical and four process corners are shown in Figure 5-16, showing that the presence of DOC does not degrade the robustness of the OpAmp. In typical case, the midband gain is 65 dB while the dc attenuation is –20 dB. The lower –3-dB frequency is 10 kHz while the unity-gain frequency is 372 MHz. The low-frequency region shows an unconditional stable response, whereas the high-frequency region shows a phase margin of ~50°. The power consumption is 2.2 mW at 1 V.

The OpAmp's flicker noise is suppressed by 33 dB (at 1 Hz) with the DOC enabled (Figure 5-17(a)). No matter the DOC is enabled or not, the white noise was measured to be –45 dBm (at 100 kHz), verifying that the DOC can suppress the $1/f$ noise without increasing the white noise in the passband. Figure 5-17(b) shows the common-mode rejection ratio (CMRR). Again, this is of no difference in the passband, it is over 110 dB at 10 kHz. The power-supply sensitivity (PSS^+ and PSS^-) with and without the DOC are less than –150 dB (Figure 5-17(c)).

Table 5-1. Sizes of the components employed in the OpAmp and DOC loop

Symbol	Size (μm)	Symbol	Size (μm)
$M_{b,1}$	2,000/0.7	$M_{oc,5-8}$	320/2
$M_{b,2-b,3}$	960/0.7	$M_{oc,9-12}$	8/14
$M_{1-2,7-8}$	1,200/0.35	$M_{oc,13-16}$	2/320
M_{3A-4A}, M_{3B-4B}	400/1.4	$M_{oc,17-18}$	320/4
M_{5-6}	80/0.85	$R_{oc,1-2}$	40 kΩ
M_{9-10}	600/0.35	R_c	0.6 kΩ
$M_{oc,1}$, $M_{oc,3}$	320/4	C_c	1.8 pF
$M_{oc,2}$, $M_{oc,4}$	100/4	C_{cp}	3 pF

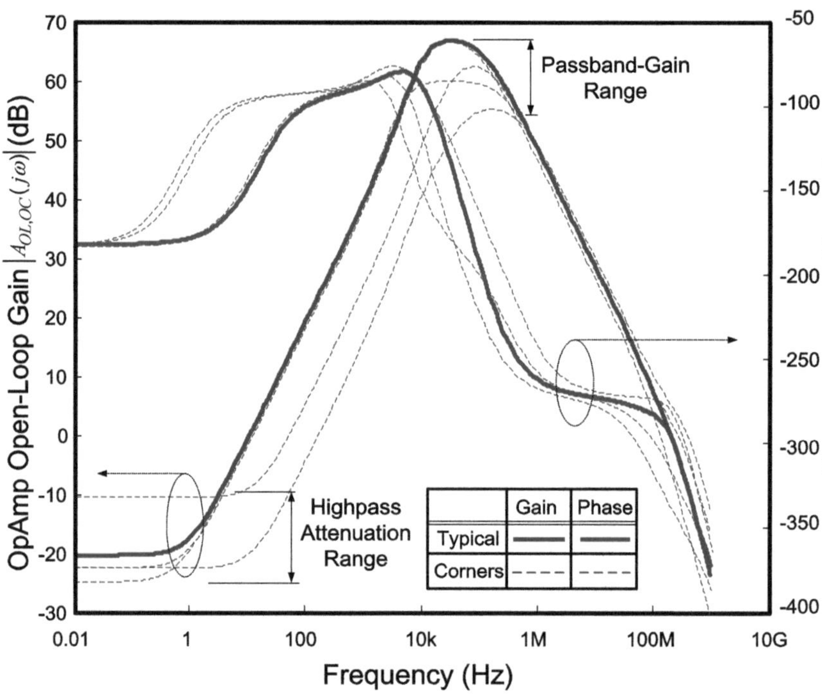

Figure 5-16. Typical and corner AC performances of the OpAmp with DOC

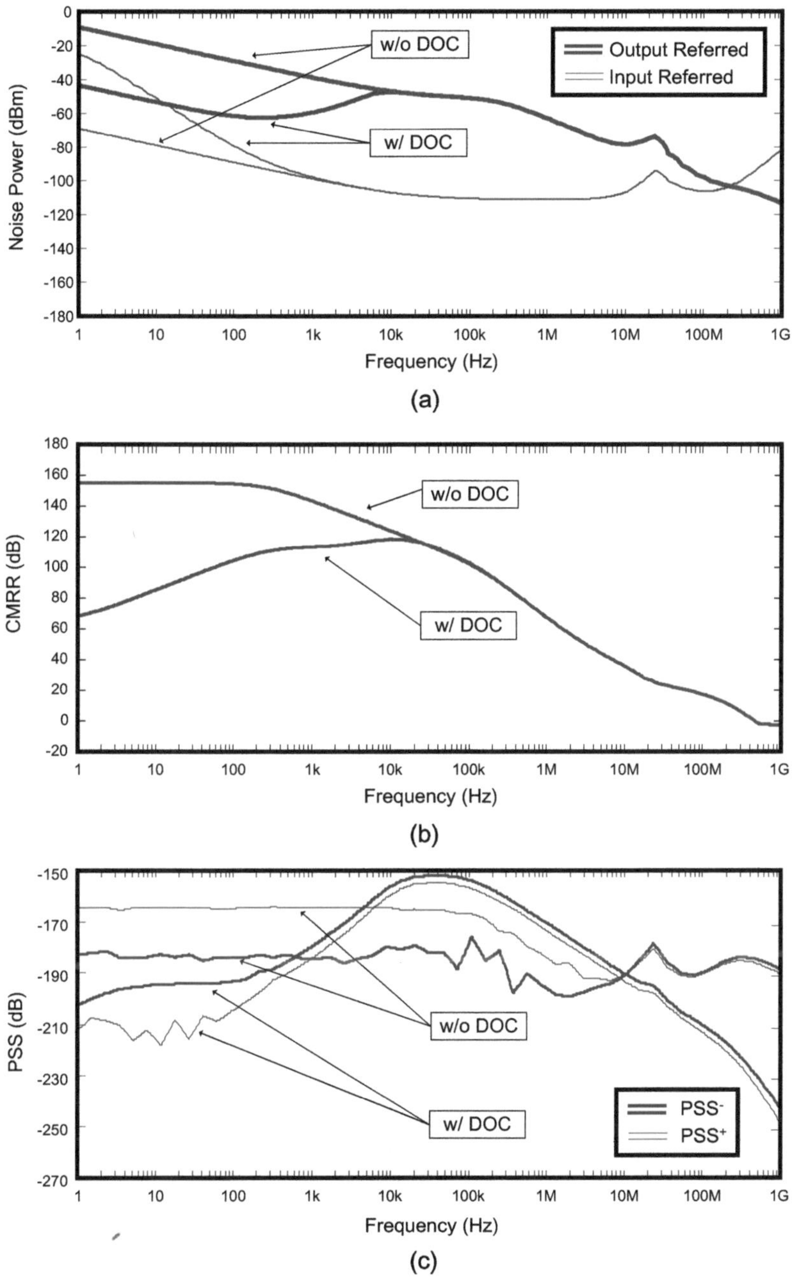

Figure 5-17. OpAmp's performances with and without the DOC enabled: (a) input-/output-referred noise (b) CMRR (c) PSS^+ and PSS^-

The histograms depicted in Figure 5-18(a)–(d) show the Monte-Carlo simulations of the unity-gain frequency, mid-band gain, phase margin and differential offset, respectively. Whereas the scatter plots depicted in Figure 5-19(a)–(c) show their correlations, verifying the robustness of the OpAmp.

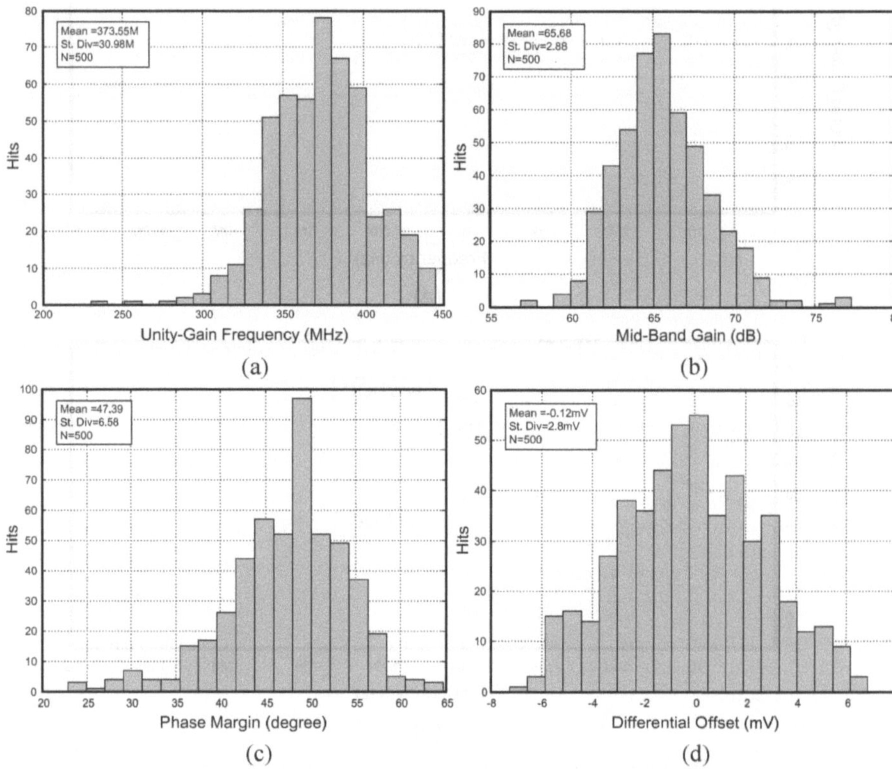

Figure 5-18. Histogram of a 500-run Monte-Carlo simulations to process variation and mismatch (No. of bins = 20): (a) Unity-gain frequency (b) Mid-band gain (c) Phase margin (d) Differential offset

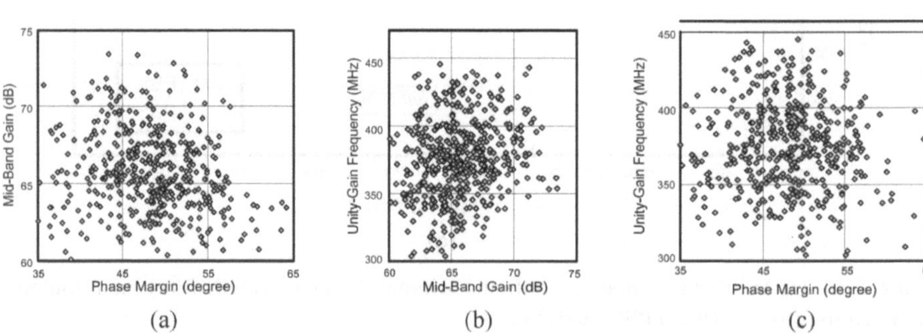

Figure 5-19. Scatter plot of a 500-run Monte-Carlo simulation to process and mismatch variation: (a) mid-band gain versus phase margin (b) unity-gain frequency versus mid-band gain (c) unity-gain frequency versus phase margin

9. SERIES-SWITCHING MIXER-QUAD, MULTI-PHASE I/Q GENERATOR AND CSF (CODESIGN)

9.1 Basic principles of switching mixer

Figure 5-20(a) and (b) show, respectively, typical current-mode switching mixers in single- and double-balanced structures. They have much better $1/f$ noise performance than their active counterparts. The input signal is mixed with a squared-wave LO signal featuring rail-to-rail amplitudes. Assuming ideal switches with zero on-resistance, the O/P voltage $V_{O/P}$, up to first harmonic, of the former is given by,

$$V_{O/P} = \frac{\delta}{2}\frac{R_{fb}}{R_{ff}}\sin(\omega_{in}t) + \frac{R_{fb}}{R_{ff}}\frac{1}{2\pi}\sin(\pi\delta)\left[\sin(\omega_{in} - \omega_{LO}t) + \sin(\omega_{in} + \omega_{LO}t)\right] + \ldots, \quad (5.10)$$

and for the latter is,

$$V_{O/P} = \frac{R_{fb}}{R_{ff}}\frac{2}{\pi}\sin(\pi\delta)\left[\sin(\omega_{in} - \omega_{LO}t) + \sin(\omega_{in} + \omega_{LO}t)\right] + \ldots, \quad (5.11)$$

where δ is the duty cycle of the LO signal (e.g., 0.5 for 50% duty cycle). ω_{in} and ω_{LO} are the angular frequencies of the input voltage (a sine wave with peak value of 1) and LO signal, respectively. The term of frequency difference is the one desired for downconversion, whereas the term of frequency summation is useful for upconversion. It is noteworthy that the latter requires differential implementation (i.e., double power and area) but it outperforms the former in conversion gain (4× larger, 2× is due to differential implementation, another 2× is obtained by swapping the differential terminals) and the I/P signal is canceled at the O/P.

9.2 Low-voltage switching mixer

For low-voltage operation, the switching mixer shown in Figure 5-20(a) can be transformed into the ones that are depicted in Figure 5-21(a) and (b), respectively. A p-type OpAmp is still assumed. Capacitors C_{lp} and C_{fb} are exploited to minimize the high-frequency mixing products in down-conversion (should be inductors for upconversion). The Case-I shown in Figure 5-21(a) employs a NMOS switch S_w in series with R_{ff}, which features a resistance much greater than that of S_w in on-state. It is noteworthy that in an n-well standard digital process, NMOS S_w suffers from body effect. Thus, the overdrive voltage (limited by V_{DD}) can be increased by using PMOS S_w with body-biasing to reduce the effective threshold voltage. The corresponding modifications $I_{b,x}$ becomes a current source from V_{DD}, and the OpAmp becomes a n-type.

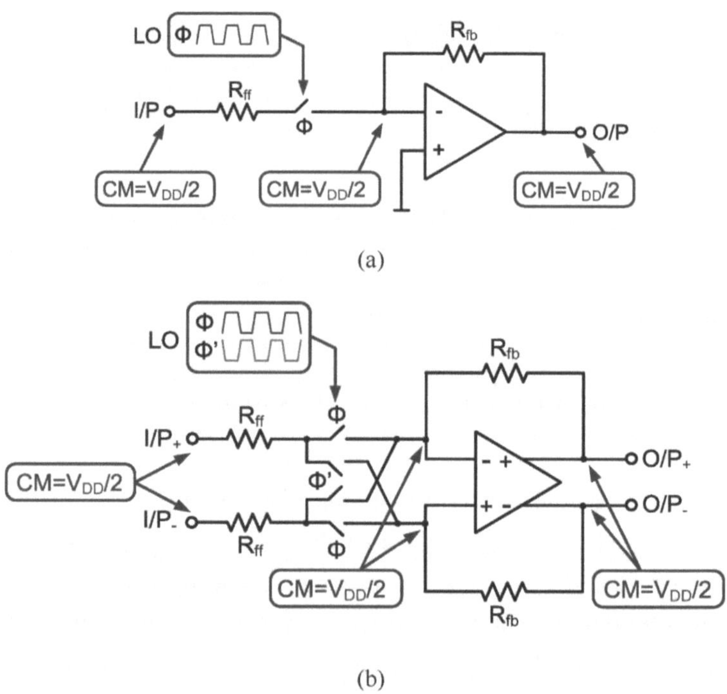

Figure 5-20. Current-mode switching mixers: (a) single- and (b) double-balanced structures

Since the dc current from the I/P is terminated periodically by the LO signal, a capacitive coupling is required to prevent altering the output CM level of the previous stage. However, capacitive coupling is in some cases unacceptable such as a direct downconversion of a narrowband channel to dc. The Case-II shown in Figure 5-21(b) is an alternative, where R_x lowers the CM level at V_z to match V_x, and $I_{b,x}$ tracks out the dc current from O/P such that no dynamic dc current exists. This circuit, however, suffers from a longer settling time and a smaller BW due to a reduced feedback factor. To eliminate the problem, a differential implementation can be used as the swapper operates with no dynamic dc current.

To ensure PVT insensitive, $I_{b,x}$ is required to match with R_{fb}, requesting a R-to-I converter in single-ended implementation. For the differential case, a CMFB is required instead.

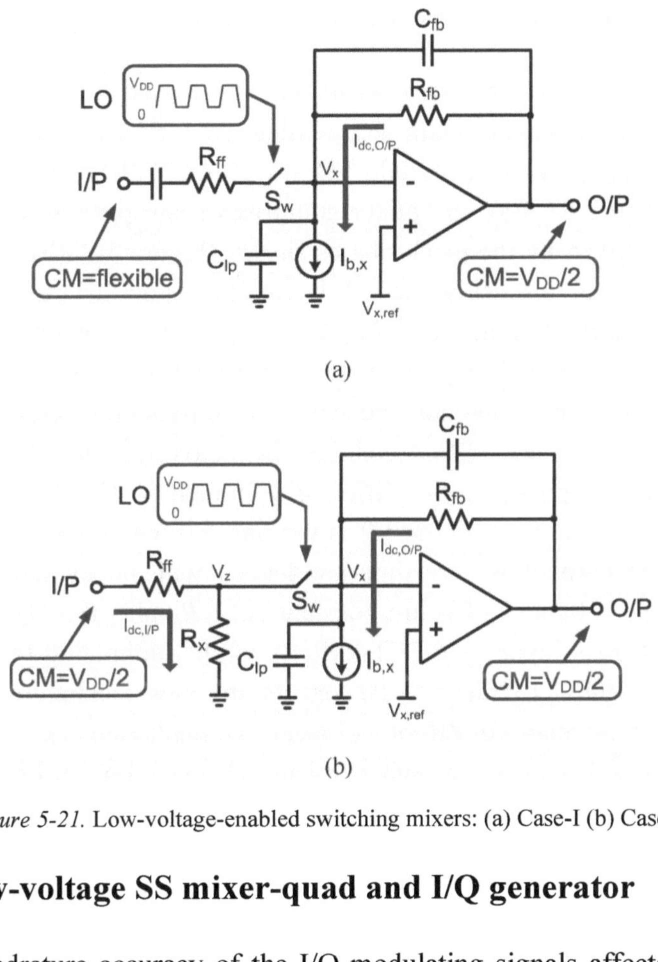

Figure 5-21. Low-voltage-enabled switching mixers: (a) Case-I (b) Case-II

9.3 Low-voltage SS mixer-quad and I/Q generator

The quadrature accuracy of the I/Q modulating signals affects critically the amount of image rejection that can be attained in the down/upconversion. In this work, a series-switching (SS) mixer-quad is proposed. It incorporates with a simple clock generator (CLKGEN) realizing a low-voltage and mismatch-insensitive I/Q downconversion.

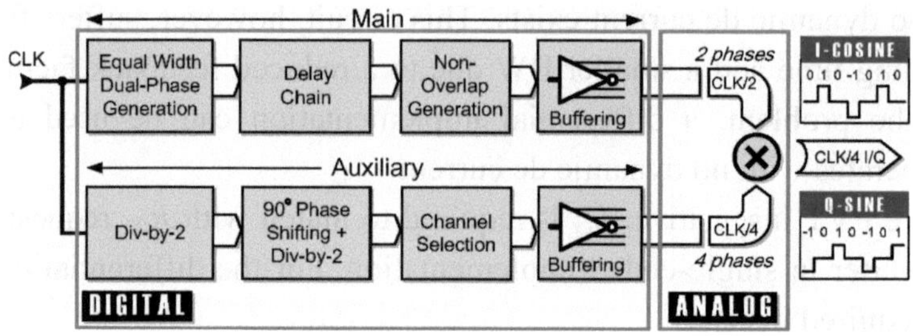

Figure 5-22. Principle of the proposed I/Q generation method

Figure 5-22 shows the basic principle; the digital part generates two mainstream clock phases, main and auxiliary, which are multiplied in the analog domain to obtain a pseudo-I/Q waveform. For the analog part, as shown in Figure 5-23(a), the mixer comprises a swapper in series with a FET-switch driven by the clock phases shown (Figure 5-23(b)). The quasi-I/Q waveform is insensitive to PVT variation due to their rail-to-rail amplitudes and the timing error (T_{AE1} and T_{AE2}) that can be tolerated. Since the swapper is activated only when the FET-switch is in open state, it does not impose any charge injection and avoids self-mixing in overall. The reset-switch driven by PH_A' (PH_B') reduces the conversion loss and memory effect during swapping of the differential branches. The mixing product from the first harmonic and the I/P is the one desired (Figure 5-24). Since the pseudo-I/Q waveform are time-interleaved with no overlap, the input impedance at $+V_{I/P}$ and $-V_{I/P}$ are constant (i.e., R_{ff} plus the equivalent on-resistance of the swapper and FET-switch), while minimizing I/Q coupling. Mismatch in the CLK-duty cycle results in only quadrature amplitude mismatch but not phase. In ZIF or LIF receivers, the harmonics other than the fundamental will translate the other in-band channels on top of the desired one, demanding a preselect filter. Alternatively, the strong third and fifth harmonics can be suppressed by using the harmonic-rejection mixer [5.18], but with passive implementation to comply with a low-voltage supply.

The implementation of the digital part can be obtained with a simple schematic (Figure 5-25). The main phases (CLK/2) are generated by the D-FFs (*D1* and *D2*) and a pair of nonoverlapping clock with a matched duty cycle. The 90°-phase shift in the auxiliary clocks (CLK/4) is obtained from the D-FFs (*D4* and *D5*) and with an inverter *N1* applied at the carrier of *D5*. It is noted that *N1* induces minor phase error as the auxiliary clock swap the inverse terminals recursively at the zero-crossings. A global set/reset initializes all flip-flops at startup.

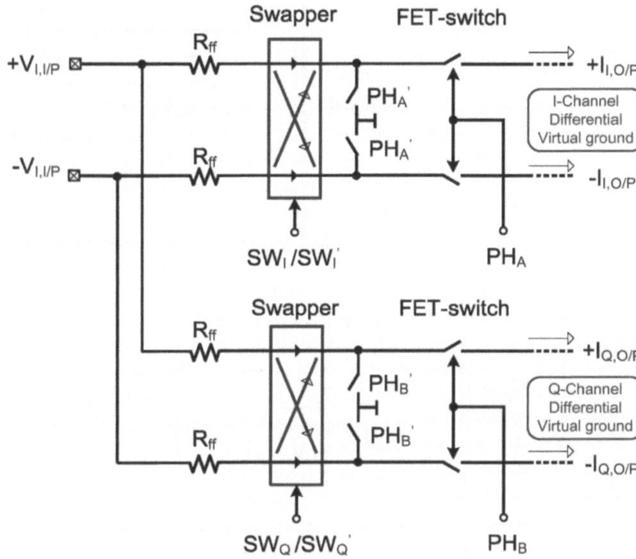

(a)

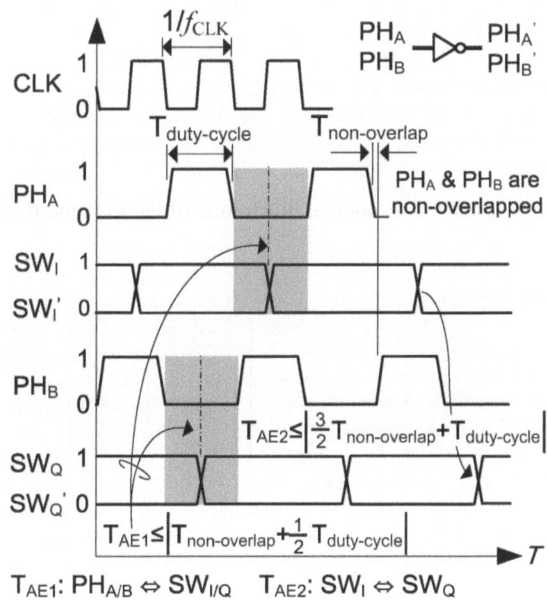

(b)

Figure 5-23. Analog part of the I/Q generation: (a) SS mixer-quad (b) and its clock phases

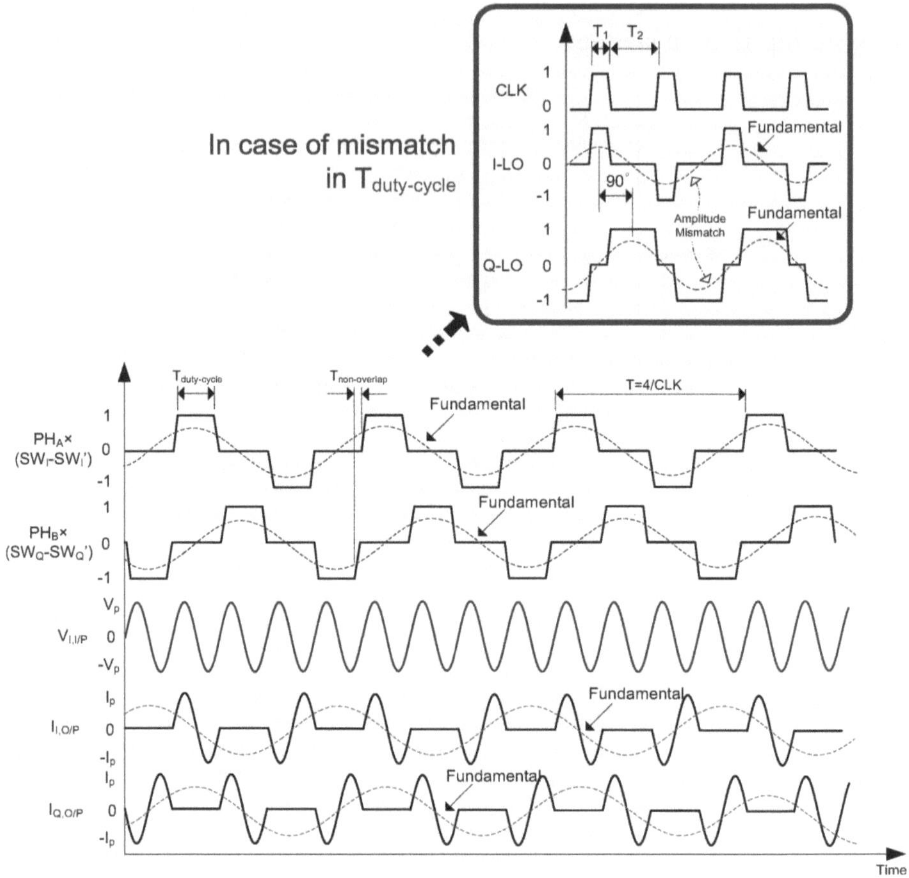

Figure 5-24. Time-domain illustration of the I/Q modulation signals

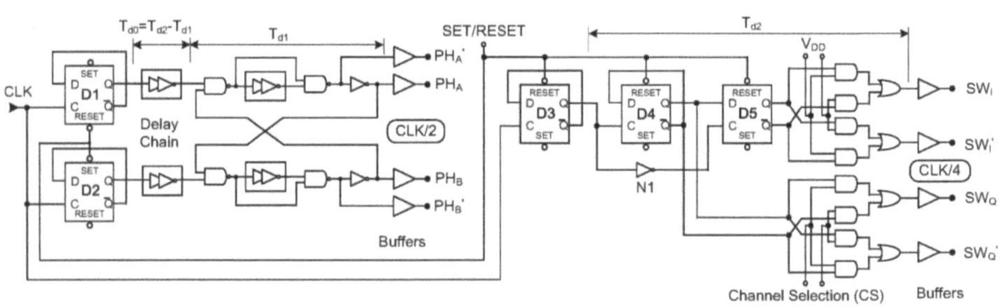

Figure 5-25. Digital part of the I/Q generation

9.4 Mismatch analysis

The image-rejection ratio (*IRR*) of the SS mixer is different from the conventional quadrature mixer [5.19] because the *IRR* of each harmonic degrades harmonically related to the phase mismatch but not with the gain. It can be shown that by assuming gain (α, a relative value) and phase (ε, in degree) mismatches the $LO_I(t)$ and $LO_Q(t)$ waveforms can be written as,

$$LO_I(t) = \frac{4A(1+\alpha)}{\pi} \sum_{n=2k+1}^{\infty} \frac{(-1)^{k+2}}{n} \cos\left(n\frac{\pi t}{T}\right) \cos(n\omega t) \quad k = 0,1,2... , \quad (5.12)$$

and

$$LO_Q(t) = \frac{4A}{\pi} \sum_{n=2k+1}^{\infty} \frac{1}{n} \cos\left(n\frac{\pi t}{T}\right) \sin(n\omega t + n\varepsilon) \quad k = 0,1,2... , \quad (5.13)$$

where n is the number of harmonic and A is the peak amplitude. The complex-sum of (5.12) and (5.13), $LO(t)=LO_I(t)+jLO_Q(t)$, is given by,

$$\begin{aligned} LO(t) = {} & \frac{4A}{\pi} \sum_{n=2k+1}^{\infty} \frac{1}{n} \cos\left(n\frac{\pi t}{T}\right) \cdot \\ & \left[(-1)^{k+2}(1+\alpha)\cos(n\omega t) + j\sin(n\omega t + n\varepsilon)\right] \quad k = 0,1,2... \end{aligned} \quad . \quad (5.14)$$

Taking the power ratio of each harmonic (H_n) to its image (I_n), the IRR_n is obtained as,

$$IRR_n = \frac{H_n}{I_n} = 10 \cdot \log\left(\frac{1 + 2(1+\alpha)\cos(n\varepsilon) + (1+\alpha)^2}{1 - 2(1+\alpha)\cos(n\varepsilon) + (1+\alpha)^2}\right) \quad n = 1,3,5,.... \quad (5.15)$$

Assuming a lower-side injection, the waveform of $LO(t)$ with 1% gain and 0.5°-phase mismatches is shown in Figure 5-26(a). The harmonics appear to interleave the positive and negative frequency axes. Mapping the result to Figure 5-26(b), the IRR_1 is ~44 dB, which is harmonically related with the phase.

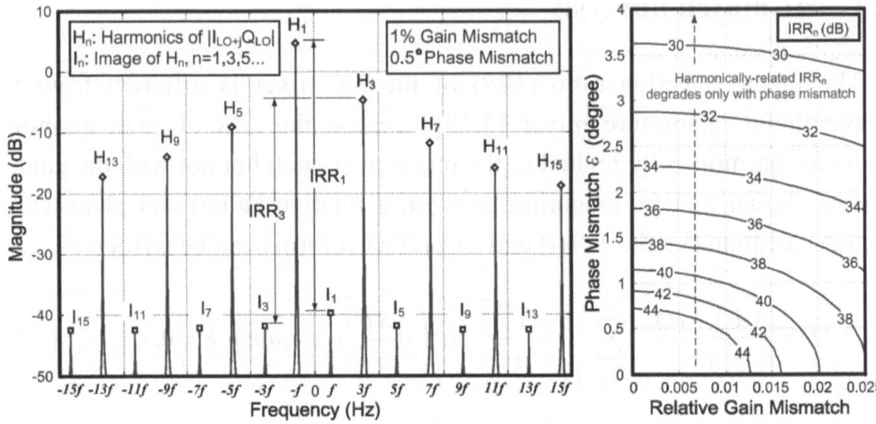

(a) (b)

Figure 5-26. (a) Spectrum of $LO(t)$ with gain and phase mismatches (b) IRR plot with gain and phase mismatches

9.5 Low-voltage CSF

The active-*RC* structure is commonly employed for low-voltage, highly linear and bandwidth-tunable CSF design. A high-order CSF is built with multiple uniquads and biquads in cascade. Figure 5-27(a) and (b) show, respectively, a kind of acitve-*RC* uniquad and biquad, with their corresponding transfer functions given by,

$$H(s)=\frac{-\dfrac{R_{fb}}{R_{ff}}}{1+sC_{fb}(R_{BW}\dfrac{R_{fb}}{R_{ff}}+R_{BW}+R_{fb})}, \qquad (5.16)$$

and

$$H(s)=\frac{-\dfrac{R_{fb}}{R_{ff}}}{1+sC_{fb}(R_{BW}\dfrac{R_{fb}}{R_{ff}}+R_{BW}+R_{fb})+s^2C_{fb}C_fR_{BW}R_{fb}}. \qquad (5.17)$$

Both are single OpAmp structures and allow an independent gain-BW control. The R_{BW} in the uniquad can be a switched resistor bank for BW tuning. Figure 5-28 shows an example, i.e., a digitally tuned resistor bank using a 7-bit control word. For the biquad, instead of tuning R_{BW} that will alter the Q factor, C_{fb} and C_f can be adjusted jointly for BW control [5.20]. Embedding the SS mixer function inside the CSF is accomplished by replacing the R_{ff} in Figure 5-27 with the circuit shown in Figure 5-23(a).

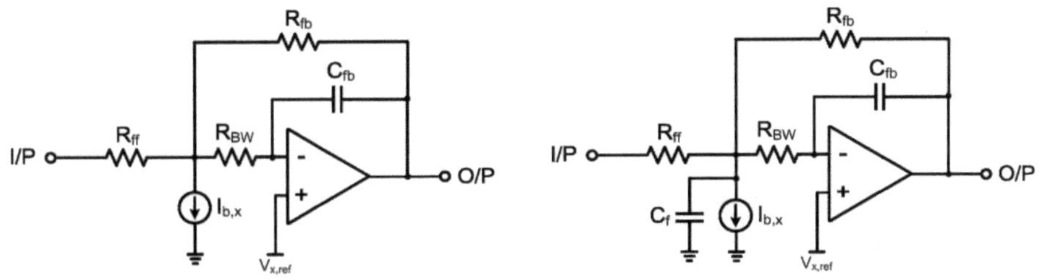

(a) (b)

Figure 5-27. Low-voltage active-RC filter: (a) uniquad (b) biquad

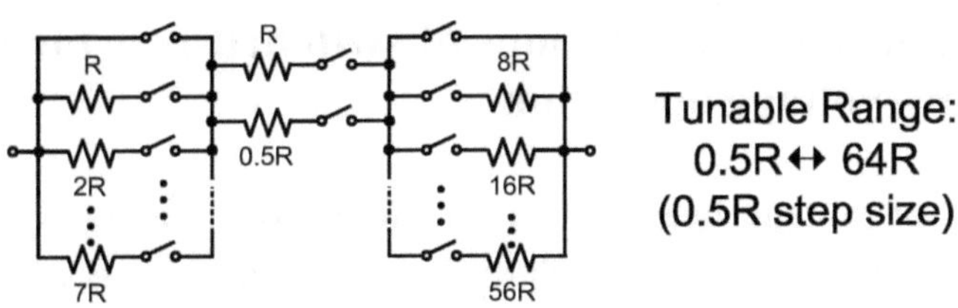

Figure 5-28. A linear-increment resistor bank for R_{BW} using a 7-bit control word

9.6 Design example and simulation results

Based on the SS mixer and mixed-mode I/Q generator, a 10-MHz double-quadrature downconverter with dual first-order passive-*RC* preselect LPFs and third-order Butterworth LPFs has been designed (system-level consideration involved). Figure 5-29 shows the simulated transient I/Q output at 8 MHz, which is a filtered result of an 18-MHz I/Q input mixed with a 10-MHz pseudo-I/Q waveform.

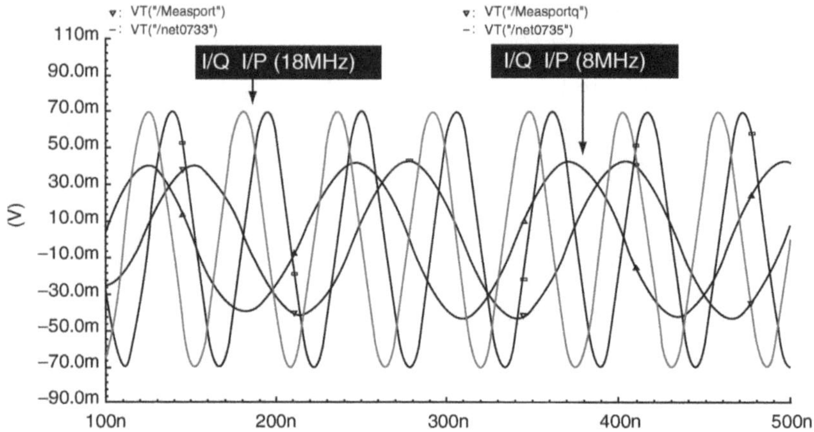

Figure 5-29. Transient I/Q waveforms of input and its corresponding outputs

10. SCR PROGRAMMABLE-GAIN AMPLIFIER

10.1 Background – limitations of switched-resistor PGA in low-voltage operation

In terms of voltage headroom, technology downscaling within the sub-micron scales brings out not much difference for designing analog blocks as along as the standard power supply V_{DD} is accompanied. For instance, an inverting amplifier using a switched-resistor gain control becomes a PGA [5.21]. However, paving the way to the sub-1-V nano-scale processes, analog circuits should be operational underneath minimum voltage headroom, rendering the classical implementation of inverting-amplifier-based switched-resistor PGA no longer effective, as discussed next.

One way to befit an inverting amplifier for a minimum V_{DD} is using a level shifter. As depicted in Figure 5-30, an extra input common-mode feedback (I-CMFB) explicitly biases the virtual ground (VG+ and VG–) to a common-mode voltage $V_{cm,in}$, that is the saturation voltage $V_{DS,sat}$ (i.e., 0.1 V) of a transistor. In our selected process, the lowest possible V_{DD} for such a PMOS differential pair becomes 1 V (i.e., $V_{DD} \geq |V_{T,p}| + 2V_{SDsat} + V_{DSsat}$). The second stage, typically a class-A amplifier, is to deliver rail-to-rail output swing by explicitly locking the output common- mode voltage $V_{cm,out}$ to $V_{DD}/2$. A large voltage swing, however, restricts the output common-mode feedback (O-CMFB) to use a resistive detector along with a current amplifier to complete the control loop. Gain tuning can be attained via replacing either the feed-forward R_{ff} or feedback R_{fb} resistor by a switched-resistor bank. The associated switch devices have to be realized with NMOS transistors and be placed at VG+ and VG– to gain an enough overdrive voltage (V_{OD}) of roughly 0.3 V (i.e., $V_{DD}-V_{T,n}-V_{DS,sat}$). Two distinct reference voltages, $V_{ref,in}$ (0.1 V) and $V_{ref,out}$ (0.5 V), are required. $V_{ref,out}$ should be a buffered one to drive the O-CMFB that draws static current.

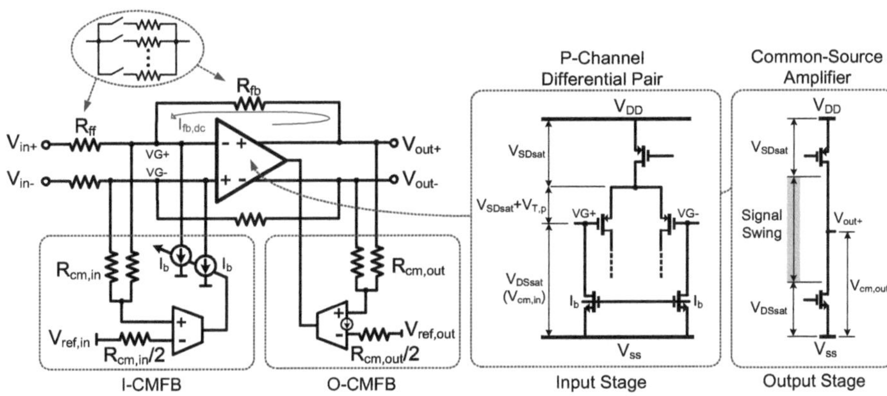

Figure 5-30. LV switched-resistor PGA modified from biased inverting amplifier

This PGA structure is LV compliant, but suffers from two drawbacks. First, independent of the V_{DD}, gain tuning through either R_{ff} or R_{fb} will vary the feedback factor, resulting in a gain-dependent output BW. Second, since the PGA's input impedance is mainly governed by R_{ff}, tuning R_{ff} without using a preceding buffer will draw a gain-dependent current from the previous stage that can be a mixer or a passive filter in a receiver. To avoid buffers for generality while facilitating the concern of loading effects in designing a multistage PGA, R_{fb} can be tuned instead. The unequal common-mode levels between $V_{cm,out}$ and $V_{cm,in}$, however, induce another gain-dependent dc current $I_{fb,dc}$ (i.e., $I_{fb,dc} = (V_{cm,out}-V_{cm,in})/R_{fb}$) in its feedback resistors, entailing a long settling time to restable the I/O-CMFBs and OpAmp at a new quiescent-operating point.

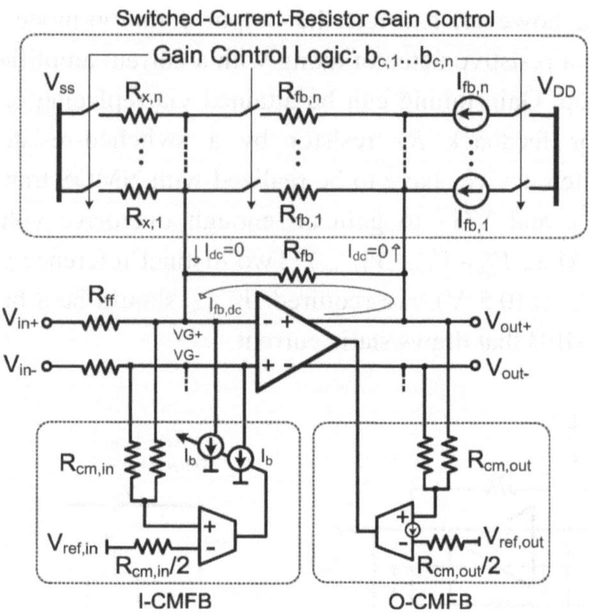

Figure 5-31. Proposed LV SCR PGA (negative terminal is omitted for clarity)

10.2 Operating principles

Illustrated in Figure 5-31 is the proposed SCR PGA for a transient-free gain control. A set of switched resistors $[R_{fb,1} \cdots R_{fb,n}]$ are added in parallel with R_{fb} to achieve a tunable gain range between the maximum $-R_{fb}/R_{ff}$ and the minimum $-(R_{fb}//R_{fb,1} \cdots //R_{fb,n})/R_{ff}$. In operation, when $[R_{fb,1} \cdots R_{fb,n}]$ are switched by the gain-control logic $[b_{c,1} \cdots b_{c,n}]$, a set of switched current sources $[I_{fb,1} \cdots I_{fb,n}]$ and grounded resistors $[R_{x,1} \cdots R_{x,n}]$ are switched correspondingly, such that $[I_{fb,1} \cdots I_{fb,n}]$ can replace the OpAmp to deliver the transient current, while $[R_{x,1} \cdots R_{x,n}]$ can sink the same current out from VG+ as given by,

$$I_{fb,n} = \frac{V_{cm,out} - V_{cm,in}}{R_{fb,n}} = \frac{V_{cm,in}}{R_{x,n}} \text{ for } n = 1,\ 2,\ 3, \ldots\ . \tag{5.18}$$

Practically, equalizing the last two terms over process, voltage and temperature (PVT) variation is uncomplicated since $V_{cm,out}$ and $V_{cm,in}$ are mirrors of $V_{ref,out}$ and $V_{ref,in}$, respectively. They can be generated underneath one master V_{DD} (i.e., $V_{ref,out} = V_{DD}/2$ and $V_{ref,in} = V_{DD}/10$), while $R_{fb,n}$ and $R_{x,n}$ can be synthesized using the same unit resistor R_u (i.e., $\alpha_n R_u = R_{fb,n} = 4R_{x,n}$, for $n = 1, 2, 3\ldots$, where α_n are a set of integers). Any PVT variation results in common-mode disturbances on both terms. Yet, matching the first term of (5.18) to the rest involves an extra signal conversion such that the practically generated switched current sources $[I'_{fb,1}\cdots I'_{fb,n}]$ can track the PVT variation of $[R_{fb,1}\cdots R_{fb,n}]$, $[R_{x,1}\cdots R_{x,n}]$, $V_{ref,out}$ and $V_{ref,in}$. A LV resistor-to-current (*R*-to-*I*) conversion circuit serves this role is proposed next.

10.3 R-to-I conversion circuit

Figure 5-32 shows the *R*-to-*I* conversion circuit for generating $V_{ref,out}$, $V_{ref,in}$, and $[I'_{fb,1}\cdots I'_{fb,n}]$ that approach the ideal $[I_{fb,1}\cdots I_{fb,n}]$ governed by (5.18). An error amplifier A_{error} in a feedback loop tracks the absolute value of R_3 under a fixed voltage V_z. The responded reference current $I_{fb,ref}$ is therefore $\propto 1/R_3$. V_z is a mirror of V_x that is set to 0.1 V ($V_{DD}/10$), enabling A_{error} to be realized simply by a *p*-channel differential pair. The R_3-tracked $I_{fb,ref}$ is then mirrored to the switched current sources $[I'_{fb,1}\cdots I'_{fb,n}]$ through transistors M_1 to $[M_{b,1}\cdots M_{b,n}]$, which features the same ratios of $R_{fb,1}$ to $[R_{fb,1}\cdots R_{fb,n}]$. Normalizing $R_{fb,1}$ as the basic element among $[R_{fb,1}\cdots R_{fb,n}]$, $[I'_{fb,1}\cdots I'_{fb,n}]$ become a function of R_3, i.e.,

$$I'_{fb,n} = \frac{V_z}{R_3}\left(\frac{R_{fb,1}}{R_{fb,n}}\right) \text{ for } n = 1,\ 2,\ 3,\ldots\ . \tag{5.19}$$

The next step is to involve $[R_{fb,1}\cdots R_{fb,n}]$ in (5.19) such that $I'_{fb,n} \propto 1/R_{fb,n}$. Matching R_3 to $R_{fb,1}$ with $R_3 = R_{fb,1}/4$ simultaneously meets the goal and equalizes the numerator of (5.19) to that of the second term in (5.18), i.e., $4V_z = V_{cm,out} - V_{cm,in}$, yielding,

$$I'_{fb,n} = \frac{4V_z}{R_{fb,n}} \text{ for } n = 1,\ 2,\ 3,\ldots\ . \tag{5.20}$$

Substituting (5.20) back to the first term of (5.18), and replacing $R_{fb,n}$ and $R_{x,n}$ according to $\alpha_n R_u = R_{fb,n} = 4R_{x,n}$, the practical expression of (5.18) is obtained, i.e.,

$$\frac{4V_z}{\alpha_n R_u} = \frac{V_{cm,out} - V_{cm,in}}{\alpha_n R_u} = \frac{V_{cm,in}}{\frac{\alpha_n}{4} R_u} \text{ for } n = 1,\ 2,\ 3,\ldots\ , \tag{5.21}$$

Recalling that V_z, $V_{cm,out}$ and $V_{cm,in}$ are mirrors of V_x ($V_{DD}/10$), $V_{ref,out}$ ($V_{DD}/2$) and $V_{ref,in}$ ($V_{DD}/10$), respectively. Any error voltage (V_Δ) on V_{DD} and error resistance (R_Δ) on R_u result no effect on the balancing of (5.21) as given by,

$$\frac{4\frac{(V_{DD} \pm V_\Delta)}{10}}{\alpha_n (R_u \pm R_\Delta)} = \frac{\frac{(V_{DD} \pm V_\Delta)}{2} - \frac{(V_{DD} \pm V_\Delta)}{10}}{\alpha_n (R_u \pm R_\Delta)} = \frac{\frac{(V_{DD} \pm V_\Delta)}{10}}{\frac{\alpha_n}{4}(R_u \pm R_\Delta)}\ , \tag{5.22}$$

yielding in overall a PVT-insensitive operation.

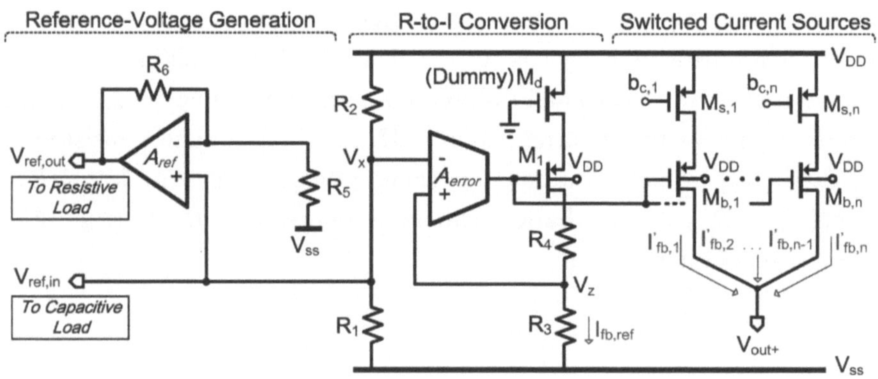

Figure 5-32. *R*-to-*I* conversion circuit for reference-voltage and switched-current-source generation

The static and dynamic performances of the SCR technique are further improved by applying the following circuit practices: (1) the current mirroring, M_1 to $[M_{b,1} \cdots M_{b,n}]$, is improved in precision by adding R_4 with $R_4 = R_3(V_D - V_z)/V_z$, level-shifting the drain voltage (V_D) of M_1 to that of $[M_{b,1} \cdots M_{b,n}]$; (2) The overall resistor matching, and the ground-noise rejection of A_{error} and A_{ref} (A_{ref} is to form a non-inverting amplifier for buffering $V_{ref,out}$), are made better by selecting $R_1 = R_2/9 = R_3 = R_4/4 = R_5 = R_6/4$, limited the resistor spread to 9; (3) $[I'_{fb,1} \cdots I'_{fb,n}]$ are switched through $[M_{s,1} \cdots M_{s,n}]$ rather than $[M_{b,1} \cdots M_{b,n}]$ such that $[M_{s,1} \cdots M_{s,n}]$ obtain the maximum overdrive voltage, leading to reduced device sizes and thereby charge injection. Moreover, since only the current paths are opened, the gate-to-source capacitance of $[M_{b,1} \cdots M_{b,n}]$ are kept charged for a faster turn-on time; (4) Connecting the bodies of $[M_{b,1} \cdots M_{b,n}]$ to V_{DD} prevents the charge injection of $[M_{s,1} \cdots M_{s,n}]$ from coupling to their gates through their body-to-gate capacitance, yielding in simulation 200–300% (depends on the gain step) shorter transients.

10.4 Feedback factor stabilization

The SCR technique also helps stabilizing the feedback factor β_{PGA} of the PGA against gain, yielding a constant-BW gain control. The idea is to keep the ratio of $[R_{fb,1} \cdots R_{fb,n}]$ to $[R_{x,1} \cdots R_{x,n}]$ identical to that of R_{fb} to R_{ff} in the expression of β_{PGA}:

$$\beta_{PGA} = \frac{1}{1 + \left(\frac{R_{fb} \,//\, R_{fb,1} \cdots //\, R_{fb,n}}{R_{ff} \,//\, R_{x,1} \cdots //\, R_{x,n}} \right)} \, . \tag{5.23}$$

Considering this ratio setting together with (5.18), a transient-free constant-BW gain control can be simultaneously achieved by satisfying the following two conditions:

$$\frac{R_{fb}}{R_{ff}} = \frac{R_{fb,n}}{R_{x,n}} \leq \frac{V_{cm,out} - V_{cm,in}}{V_{cm,in}} \quad \text{for } n = 1,\ 2,\ 3, \ldots\ , \tag{5.24}$$

and,

$$\beta_{PGA} \leq \frac{V_{cm,in}}{V_{cm,out}} \, . \tag{5.25}$$

As it will be presented in the next subsection via two design examples, satisfying both (5.24) and (5.25) are practically simply. One may also observe that (5.24) and (5.25) are just ratio dependent, leading to a PVT-insensitive operation. Not counting the advantage of constant BW, an unchanged β_{PGA} has the advantages of unvarying settling time and constant stopband rejection.

Continuing from previous discussion under a 1-V V_{DD}, $V_{cm,in}$ of 0.1 V and $V_{cm,out}$ of 0.5 V directly set the PGA's maximum gain and β_{PGA} to 4 (i.e., 12 dB) and 0.2, respectively. The implications to a 1-V and a sub-1-V implementation are demonstrated in the following two examples.

10.5 SCR PGA in 1-V and sub-1V operation

Two cases using distinct V_{DD}s, 1 V and 0.6 V, are presented to give an insight of the SCR technique.

1. A 1-V 24-dB-Gain-Range SCR PGA – For a 1-V V_{DD}, $V_{cm,in}$ = 0.1 V and $V_{cm,out}$=0.5 V are accordingly set to respect the aforesaid concerns. To offer, for instance, a gain range of –12 to 12 dB with a 6-dB step size, we set $R_{fb} = 4R_{ff}$ and $4R_{x,3-n} = R_{fb,3-n} = 2^n R_{ff}$ for $n = -1, 0, 1, 2$, resulting in a constant β_{PGA} of 0.2 while satisfying (1) for a transient-free gain control. Without the technique, β_{PGA} will vary between 0.2 (at 12 dB) and 0.8 (at –12 dB), equivalent to a BW difference by a factor of 4. Of course, in practice, the constancy of β_{PGA} relates to the resistance ratio of $R_{cm,in}$ to R_{ff} and $[R_{x,1} \cdots R_{x,n}]$. Under the same numerical conditions given above (5.23) can be rewritten as,

$$\beta_{PGA} = \frac{1}{5}\frac{1}{1+\frac{4}{5}\left(\frac{R_{ff} \,//\, R_{x,1} \cdots // \, R_{x,n}}{R_{cm,in}}\right)}. \tag{5.26}$$

For instance, if $R_{cm,in}$ is 5× greater than $R_{ff}//R_{x,1} \cdots //R_{x,n}$, β'_{PGA} will vary slightly between 0.177 (at 12 dB) and 0.198 (at –12 dB), resulting in a BW variation of around 12%.

2. A 0.6-V 24-dB-Gain-Range SCR PGA – If the V_{DD} is downscaled to 0.6 V, $V_{cm,in}$ of 0.1 V and $V_{cm,out}$ of 0.3 V are appropriate choices. To attain the same gain range as the former, two identical PGAs in cascade are required because the maximum closed-loop gain, governed by (5.24), is 2 (~6 dB) in magnitude. Each PGA offers a gain range of –6 to 6 dB with a 6-dB step size by setting $R_{fb} = 2R_{ff}$ and $2R_{x,2-n} = R_{fb,2-n} = 2^n R_{ff}$ for $n = 0, 1$. Comparing with the former, a transient-free gain control is still achieved but it will result in a larger β_{PGA} of 1/3. Although two identical PGAs in cascade reduce the BW by 35% (i.e., multiplied by a factor of $\sqrt{[2^{1/N}-1]}$, where N is the number of stage in cascade), the larger β_{PGA} in turn results in a 67% increment in BW if the same OpAmp specifications are met, in turn giving a net BW enlargement of 32%. Of course, two PGAs imply double of power, constituting a roughly-fair trade-off between power and V_{DD}. A BW variation similar to the former is also discarded.

10.6 Linearity consideration

For a fully differential circuit implementation with dc-offset cancellation, the even-harmonic distortion can be suppressed effectively such that only the odd harmonics are dominant. With a NMOS switch positioned in series with a feedback resistor, and assuming a sinusoidal input signal, the third-harmonic distortion (HD3) can be estimated by (similar to the analyzing method presented in [5.22]),

$$HD3 \approx \frac{3}{32}\left(\frac{V_{out-,p}}{V_g - V_{cm,in} - V_{T,n}}\right)^2 \cdot \left(\frac{r_{on}}{R_{fb}}\right)^3 , \tag{5.27}$$

where V_g is the transistor gate voltage, $V_{out-,p}$ is the peak value of the output voltage and r_{on} is the on-resistance of transistor. For example, with V_g=1 V, $V_{out-,p}$= 0.5 V, $V_{cm,in}$= 0.1 V, $V_{T,n}$= 0.52 V, and R_{fb} = 2.5 kΩ for a minimum-gain level, r_{on} can be as large as 213 Ω (8.5% of R_{fb}) for a HD3 of –80 dB. This indicates that explicitly biasing $V_{cm,in}$ to a value close to one of the supply rails not only allows a LV operation but also improves the linearity due to an increase of V_{OD}.

10.7 Noise consideration

The equivalent noise model of Figure 5-31 is presented in Figure 5-33. Since the PGA, in receiver use, is preceded by a high-gain filter, its gain-dependent input-referred noise is uncritical to the overall noise figure. In the following analysis, only thermal noise is taken into account since highpass poles are created at dc for each stage. The mean squared output noise v^2_{noise} of the SCR PGA is given by,

$$v_{noise}^2 = \left(\frac{4kT}{R_{ff}} + \frac{4kT}{R_{x,eq}} + \frac{4kT}{R_{fb,eq}} + I_{b,n}^2 \right) R_{fb,eq}^2 \\ + v_{oa,n}^2 \left(1 + \frac{R_{fb,eq}}{R_{ff} // R_{x,eq} // r_{ds} // R_{cm,in}} \right)^2, \quad (5.28)$$

where, $R_{fb,eq} = R_{fb}//(R_{fb,1}+r_{on,fb,1})\cdots//(R_{fb,n}+r_{on,fb,n})$, $R_{x,eq} = (R_{x,1}+r_{on,x,1})\cdots//(R_{x,n}+r_{on,x,n})$, k is the Boltzmann constant, T is the absolute temperature, r_{ds} is the output resistance of the current source I_b. $v_{oa,n}$ is the equivalent input-referred noise voltage of the OpAmp, which includes the uncritical noise contribution of $[I_{fb,1}\cdots I_{fb,n}]$ that are injected at the output stage {i.e., $(8/3)\times[kT/(g_{m,1}\cdots//g_{m,n})]$, where $g_{m,1}\cdots//g_{m,n}$ are the transconductances of $[M_{b,1}\cdots M_{b,n}]$} and the excess noise coefficient γ is assumed to be 2/3. $I^2_{b,n}$ is the mean squared noise current of I_b as given by,

$$I_{b,n}^2 = \frac{16}{3} kT \frac{V_{cm,out} - V_{cm,in}}{V_{cm,in}(R_{fb,eq} // R_{ff} // R_{x,eq})}, \quad (5.29)$$

which indicates that keeping the resistor spread small and increasing the level of $V_{cm,in}$ are imperative to lower v^2_{noise}, but there remains a trade-off in stage gain range and linearity.

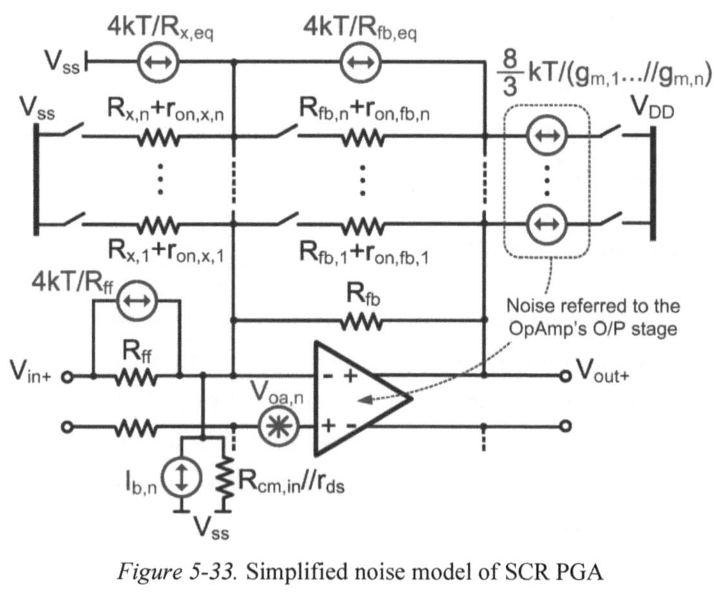

Figure 5-33. Simplified noise model of SCR PGA

10.8 Design example

To demonstrate the proposed SCR and inside-OpAmp DOC techniques, a 52-dB-gain-range 3-stage PGA was designed, which incorporates a channel-selection filter that can offer 20-dB gain to meet the baseband gain-range requirement of 802.11a/b/g (i.e., 0–50 dB). As shown in Figure 5-34, unlike the conventional servo loop that closes the feedback with multiple forward stages, here each-stage OpAmp has a local DOC to ensure a balanced internal signal transfer and facilitates the stability concerns [5.23]. Coarse (6 dB/step) followed by fine (2 dB/step) gain controls were structured to reduce the global gain-control transients [5.24]. The simplified schematics of the PGA's 1st-(2nd-) and 3rd-stage are depicted in Figure 5-35(a) and (b), respectively. Here, the values of the resistor ratios maintain the feedback factor constant at 0.2 and stabilize the quiescent-operating points of the OpAmp, I-CMFB and O-CMFB by generating $I_{fb,1-6}$ using the previously mentioned *R*-to-*I* conversion circuit. The 3rd-stage shows a small round-off error since its β_2 is deliberately reduced from the original minimum of 0.39–0.2, such that only one OpAmp design is involved in all stages. The following simulation results, presented from the OpAmp to the overall PGA, substantiate the performance claims with simple tests.

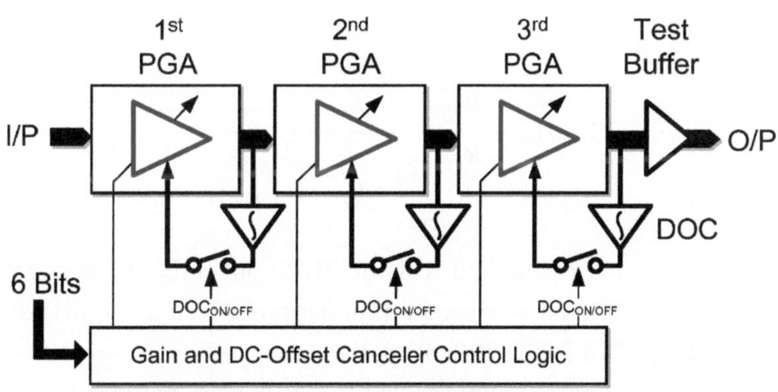

PLAN	1ST	2ND	3RD	TOTAL
MAX.	12dB	12dB	6dB	30dB
MIN.	-12dB	-12dB	2dB	-22dB
STEP	6dB	6dB	2dB	

Figure 5-34. Implemented 3-stage PGA and gain plan

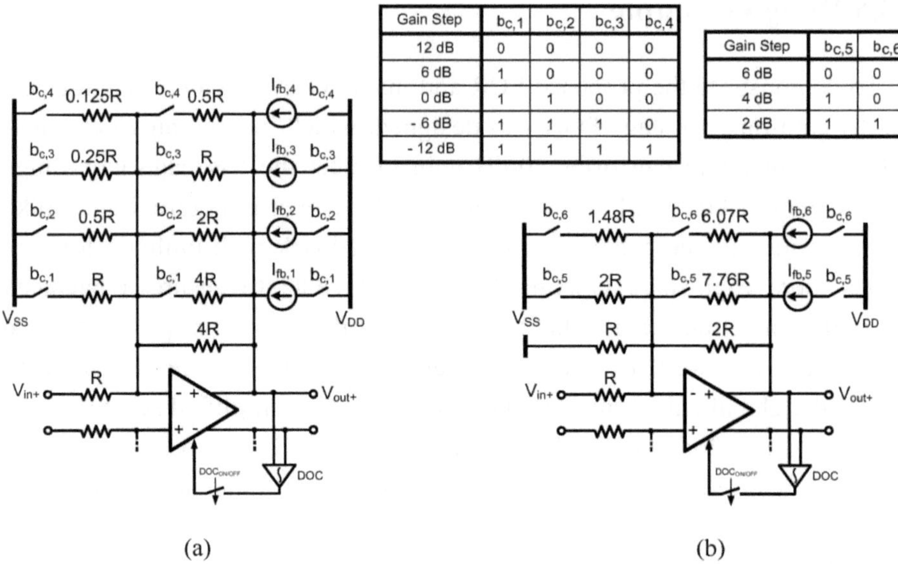

Gain Step	bc,1	bc,2	bc,3	bc,4
12 dB	0	0	0	0
6 dB	1	0	0	0
0 dB	1	1	0	0
- 6 dB	1	1	1	0
- 12 dB	1	1	1	1

Gain Step	bc,5	bc,6
6 dB	0	0
4 dB	1	0
2 dB	1	1

Figure 5-35. Simplified schematics of the PGA: (a) 1st /2nd and (b) 3rd stage

DOC in switched-resistor and SCR PGAs – employing such an OpAmp in the switched-resistor PGA and SCR PGA will lead to the closed-loop gains depicted in Figure 5-36(a) and (b), respectively. It is a high-level estimation (i.e., the phase shift at low frequency is assumed to be small) by using,

$$A_{CL,OC}(j\omega) = -\frac{R_{fb}}{R_{ff}} \frac{1}{(1+\dfrac{1}{A_{OL,OC}(j\omega)\beta})}, \qquad (5.30)$$

where β refers to β_{IA} for switched-resistor PGA; but refers to β_{PGA} for SCR PGA. The former offers an inconstant attenuation at dc (22–34 dB) due to a variation of β_{IA} (0.8–0.2). Differently, the latter offers a 34-dB constant attenuation at dc due to a constant β_{PGA} of 0.2. It implies that not just the BW is stabilized, but also the rejection at dc, giving a true dc-offset transient-free gain control. Figure 5-37(a) and (b) show how the rejection at dc being enhanced and linearized in the constant β case for the 1st (2nd) and 3rd stages, respectively.

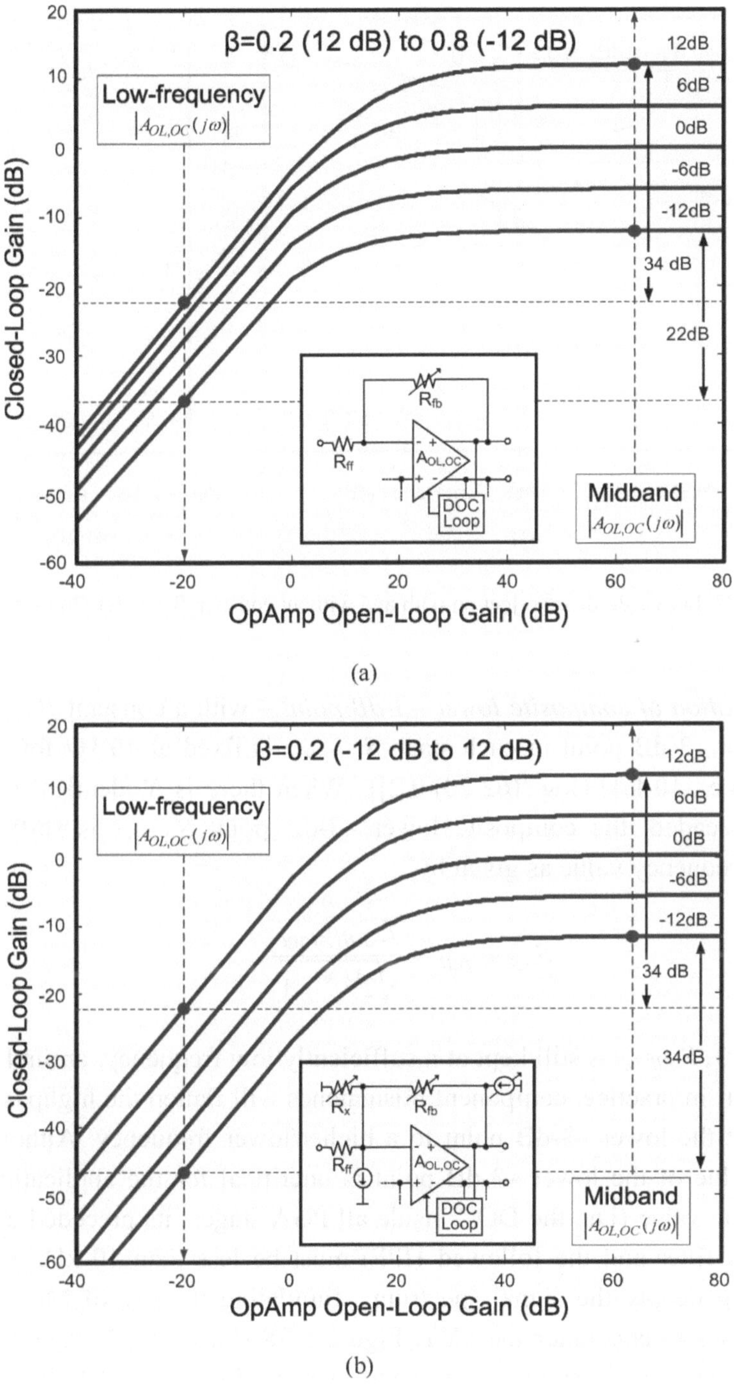

Figure 5-36. PGA's closed-loop gain versus OpAmp's open-loop gain at low-frequency and mid-band: (a) switched-resistor PGA (b) SCR PGA

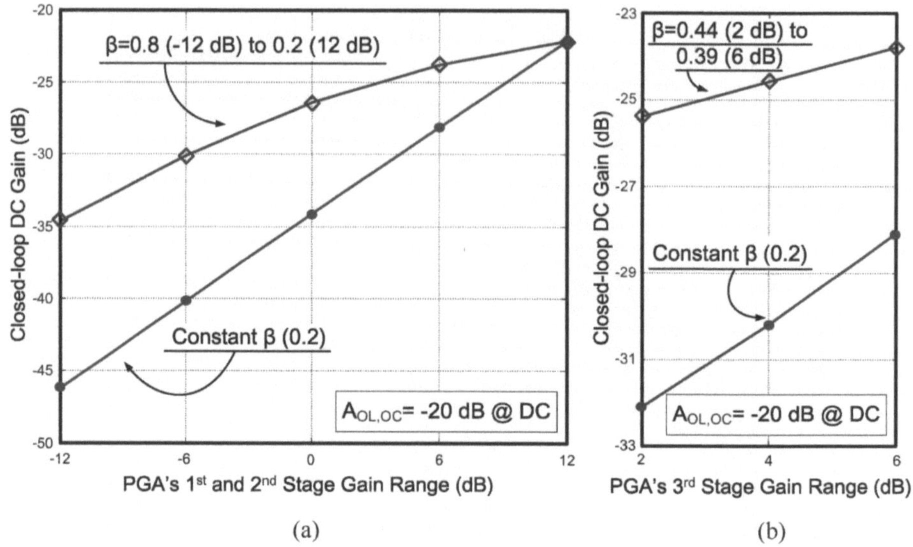

Figure 5-37. PGA's dc-offset rejection with and without constant β: (a) 1st /2nd stage (b) 3rd stage

Estimation of composite lower –3-dB point – with a constant β_{PGA} of 0.2, the lower –3-dB point in each stage $S_{-3dB,stage}$ is fixed at 40 Hz for all gain levels {i.e., 10 k/[1+log^{-1}(62/20)·0.2]}. When there is N identical highpass stage cascaded, the composite lower –3-dB point $S_{-3dB,full}$ is shifted to a higher frequency value as given by,

$$S_{-3dB,full} = \frac{f_{-3dB,stage}}{\sqrt{2^{1/N} - 1}} \quad . \tag{5.31}$$

With N=3, $S_{-3dB,full}$ is still kept at a sufficiently low frequency, around 78 Hz. However, in practice, component mismatches will flatten the highpass notch and shift the lower –3-dB point to a higher/lower frequency. Although the exact value of the lower –3-dB point is uncritical for the applications, the composite value (i.e., the DOC inside all PGA stages, its preceded channel-selection filter and the followed HPF) must be less than 10 kHz to avoid damaging deeply the signal spectrum, stimulating the use of Monte-Carlo simulations to encounter the PVT. Figure 5-38 shows the 3-stage-cascaded PGA's magnitude responses simulated over random mismatch and process variations. The lower –3-dB point is maximally less than 3 kHz while offering minimally a rejection of 65.2-dB around dc.

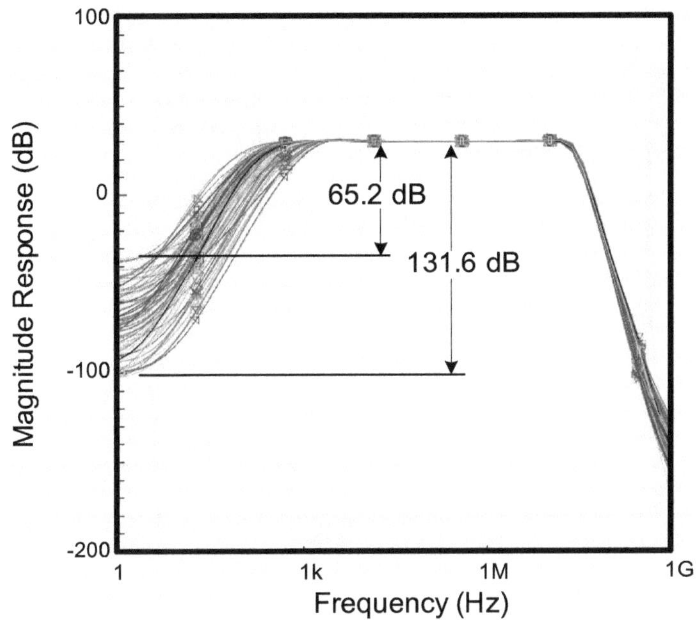

Figure 5-38. 100-time Monte-Carlo simulation results of the 3-stage PGA's magnitude response at 30-dB gain

Step response and gain control of the PGA in receiver use – to provide a fast tracking for dc-offset transient, the 3-stage PGA is followed by a first-order passive-*RC* highpass filter (HPF) in the implemented receiver. Its pole is switchable to a high/low frequency for preamble/normal reception such that the composite lower –3-dB point is 1 MHz/10 kHz, respectively. Simulated at the highest 30-dB gain level with a step input and 5% channel mismatch artificially assigned in all PGA stages, the settling time is extremely long if no switching is applied (Figure 5-39), but is shorted to less than 0.5 μs by abruptly switching off the three DOCs and shifting the pole of the HPF to high frequency at the beginning. Afterwards, the three DOCs are switched on progressively in three time slots that are synchronized with the pole switching of the HPF back to lower frequencies. The simulated transient in a 52-dB gain step is 240 ns as shown in Figure 5-40.

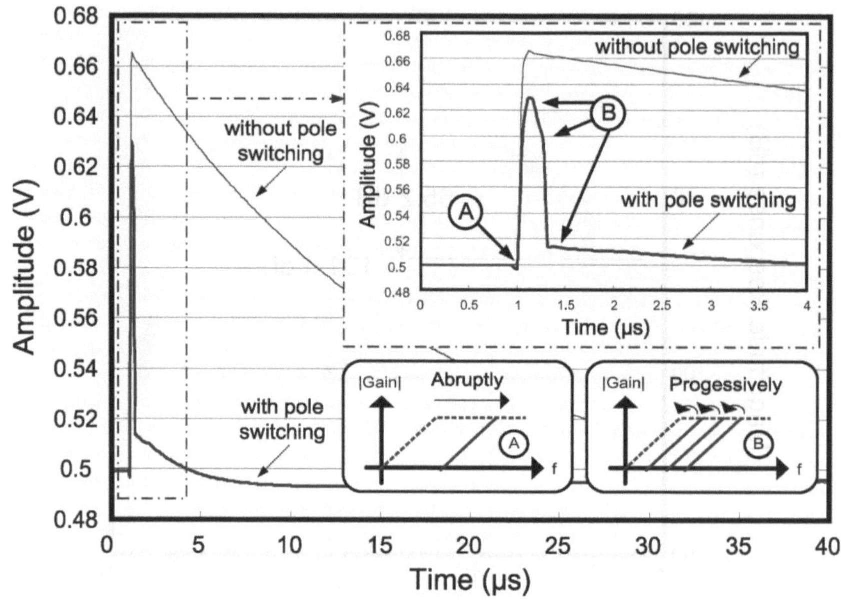

Figure 5-39. PGA step response with and without a pole-frequency control in the DOC

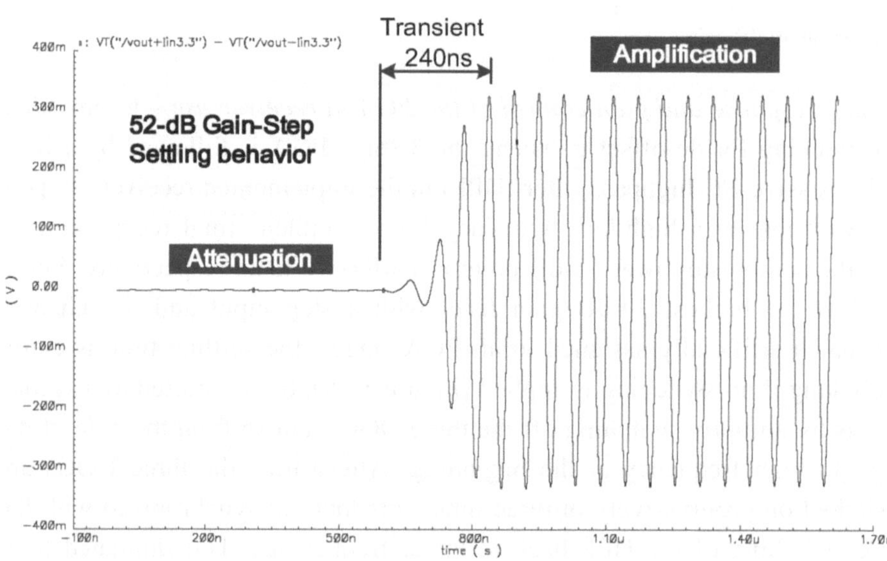

Figure 5-40. Transient simulation with a 52-dB gain step applied

Systematic dc-offset removability of the DOC – a differential ramp input can determine the systematic dc-offset removability of the PGA. Figure 5-41 plots the simulated output dc offset of a single-stage 0-dB-gain PGAs versus a differential ramp input swapped between ±0.5 V. The output follows the ramp input with the same slope when the DOC is disabled, but is suppressed notably when it is enabled. From the lower subplot of Figure 5-41, we can observe that the residual output dc offset after suppression is 2.4-/3.8-/8.9-/ 19.6-mV differential, with an input dc offset of 100-/200-/300-/400-mV differential applied. The dead zone only happens when the input is extremely small. The output within the dead zone has a slope matching the case when the DOC is disabled (upper subplot), showing also that the irremovable dc offset ($V_{in+}-V_{in-}$) is less than 5 mV.

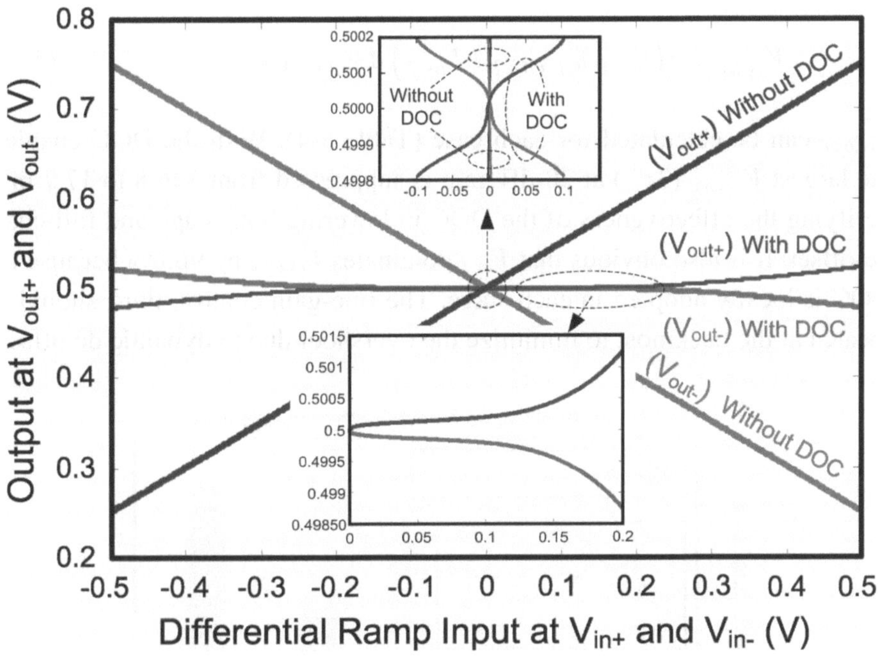

Figure 5-41. Output dc offset of a single-stage 0-dB-gain PGA versus a ramp input

Estimation of random mismatches induced dc-offset – Monte-Carlo simulation is a tool that can simultaneously take systematic and random dc offsets into account. The dc offset following a normal distribution shows a mean value of 0 and a standard deviation of σ_{os} related to the gain step. The simulated σ_{os}s of a single-stage PGA at 12, 6, 0, –6 and –12-dB gain are shown in Figure 5-42. With the DOC disabled, σ_{os} increases with the gain from 5.8 mV to 10.7 mV. Alternatively, with the DOC enabled, σ_{os} is less than 6 mV, implying 99.75% ($3\sigma_{os}$) one stage yields less than 18 mV dc offset. Based on such results, the dc offset of the PGA's 1st, 2nd and 3rd stages, $V_{os,1}$, $V_{os,2}$ and $V_{os,3}$, are accordingly obtained in Table 5-2, where only the two extreme gain levels (i.e., –22 and 30 dB) of the three stages in cascade are shown. To estimate the total output dc-offset $V_{os,total}$, the PGA's 1st-/2nd-/3rd-stage dc gain, $A_{CL,OC,1}/A_{CL,OC,2}/A_{CL,OC,3}$, with the DOC enabled and disabled are computed first, as listed in Table 5-3 (where the results obtained with the DOC enabled come from Figure 5-16). Eventually, using the output-referred dc-offset calculation model shown in Figure 5-43, and substituting the results from Table 5-2 and 5-3 to,

$$V_{os,total} = \left(V_{os,1}A_{CL,OC,2} + V_{os,2}\right)A_{CL,OC,3} + V_{os,3} \, . \qquad (5.32)$$

$V_{os,total}$ can be calculated for each case (Table 5-4). With the DOC enabled, the largest $V_{os,total}$ ($3\sigma_{os}$) at 30-dB gain is suppressed from 346.8 to 17.7 mV, verifying the effectiveness of the DOC in lowering both stage and full-chain dc-offset. It is also obvious that $V_{os,3}$ dominates $V_{os,total}$ by 96.6% because the DOC is locally adopted in each stage. The fine-gain control, thus, should be located at the backmost to minimize the overshoot due to dynamic dc offset.

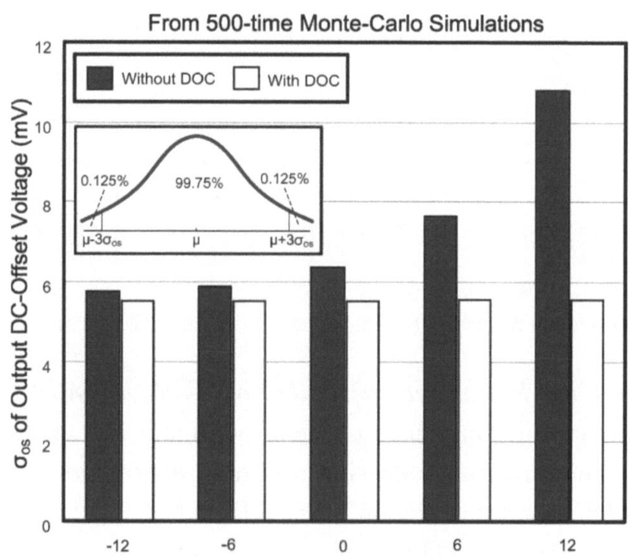

Figure 5-42. σ_{os} of a single-stage PGA's output dc-offset voltage

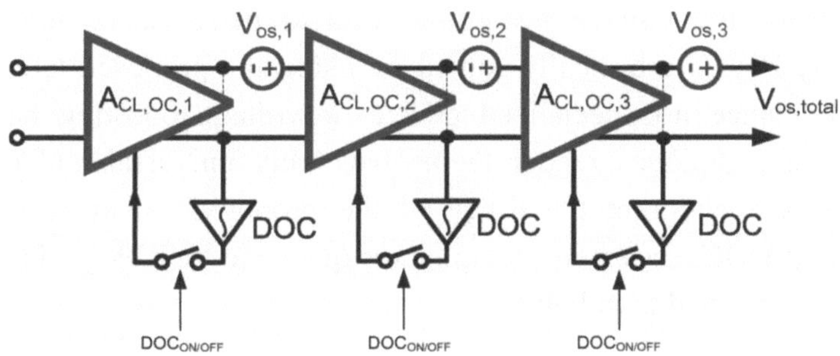

Figure 5-43. $V_{os,total}$ calculation using an output-referred approach

Table 5-2. $3\sigma_{os}$ Monte-Carlo simulated stage-output dc offset

Output gain level	Stage-output dc offset (mV) ($3\sigma_{os}$)					
	With DOC			Without DOC		
	$V_{os,1}$	$V_{os,2}$	$V_{os,3}$	$V_{os,1}$	$V_{os,2}$	$V_{os,3}$
–22 dB	16.8	16.8	16.95	17.4	17.4	20.1
30 dB	17.1	17.1	17.04	32.4	32.4	22.8

Table 5-3. Simulated stage closed-loop dc gain

Output gain level	Stage closed-loop dc gain (dB)					
	With DOC			Without DOC		
	$A_{CL,OC,1}$	$A_{CL,OC,2}$	$A_{CL,OC,3}$	$A_{CL,OC,1}$	$A_{CL,OC,2}$	$A_{CL,OC,3}$
–22 dB	–22	–22	–28.1	–12	–12	2
30 dB	–46.2	–46.2	-32.1	12	12	6

Table 5-4. Simulated total (3-Stage) output dc offset ($3\sigma_{os}$)

Output gain level	Total-output dc offset (mV) ($3\sigma_{os}$)	
	With DOC	Without DOC
	$V_{os,total}$	$V_{os,total}$
–22 dB	17.37	47.52
30 dB	17.7	346.8

11. TECHNIQUES REUSABILITY IN ADVANCED TECHNOLOGY NODES

Successful LV techniques should keep their robustness in the upcoming technology nodes. It is expected that when the technology downsizes to nanoscale, the ratio of V_{th}/V_{DD} is approaching a value around 0.5 (Chapter 1). In this research, such a ratio taken is even challenged, ranged between 0.52

and 0.65. Thus, the voltage headroom resulted performance limitations, especially the BW, are basically solved by using the proposed techniques. In particular, since no specialized device or voltage boosting has been employed, the techniques render themselves truly universal and reliable. Other remarks concerning the usage of the proposed 3 key techniques: inside-OpAmp DOC, series-switching mixer-quad and SCR PGA, in sub-1V technology nodes are drawn below.

- As the physical size of MOS transistor keeps shrinking, dc offset due to component mismatches will become harder to handle in one loop where many high-gain blocks are in cascade. The proposed inside-OpAmp DOC is an area-efficient approach to design OpAmp with built-in switchable highpass pole for a flexibly dc-offset removal. For instance, in the baseband of a receiver, all OpAmp-based circuits can use the inside-OpAmp DOC technique to maximize the signal swing and prevent the dc offset from propagation down the receiver chain.
- Although the series-switching mixer-quad operated in current mode provides no signal gain, actually loss, the linearity is more V_{DD}-independent than its voltage-mode counterpart, befitting a sub-1V supply. In addition, since such a mixer requires simple digital clock phases for I/Q generation and is highly mismatch insensitive, it can be widely exploited as an image-rejection down/upconverter.
- Finally, for the SCR PGA, further reducing the V_{DD} will require more cascaded stage to maintain the properties of transient free and constant BW, under the same programmable gain requirement. Architecturally, the trade-off involved in the SCR PGA in technology scaling is that, reducing the V_{DD} by a factor will roughly lead to an increase of power by the same factor. However, the intrinsic f_T of upcoming transistors is continuously growing up. An OpAmp with a higher gain-bandwidth product is more power efficient to achieve, gaining the speed advantage offered by technology scaling.

12. SUMMARY

This chapter has presented various circuit techniques for the realization of CT analog-baseband circuitry under LV constraints. The described blocks include, OpAmp, CT level shifter, CMFB, current switch, MOS capacitor, dc-offset cancellation circuit, IF downconverter, I/Q generator, CSF and PGA. In the simplest structures, the techniques based on inverting amplifier can be generalized in Figure 5-44. Those blocks form a technology independent portfolio for the implementation of truly LV analog circuits. Experimental results of the proposed LV circuits, and their combined receiver analog baseband, will be presented together in Chapter 6.

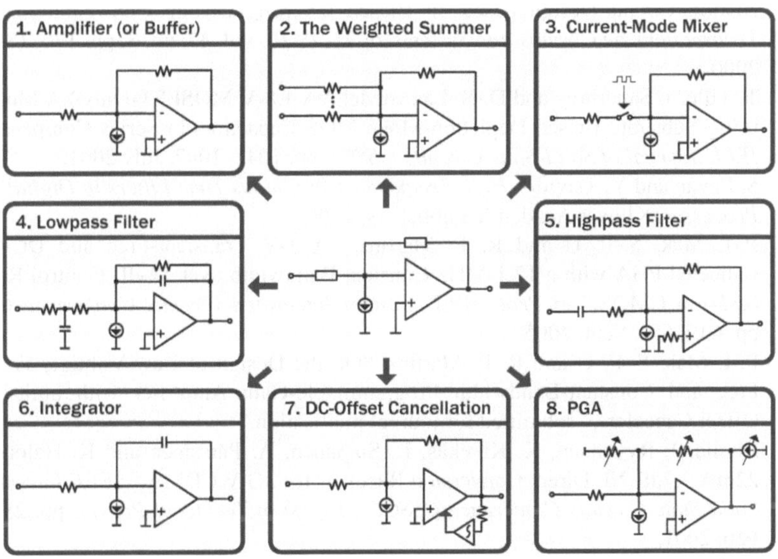

Figure 5-44. Low-voltage analog functions based on inverting configuration

REFERENCES

[5.1] L. Y. Matsuya and J. Yamada, "1V Power Supply, Low-Power Consumption A/D Conversion Technique with Swing-Suppression Noise Shaping," *IEEE Journal Solid-State Circuits (JSSC)*, vol. 29, no.12, pp. 1524–1530, Dec. 1994.

[5.2] J. R. Angulo, S. C. Choi and G. G. Altamiran, "Low Supply Voltage OTA Architectures using Floating Gate Transistors," in *Proc. IEEE Midwest Symposium on Circuits and Systems (MWSCAS)*, vol. 1, pp. 158–162, Aug. 1995.

[5.3] S. Chatterjee, Y. Tsividis and P. Kinget, "A 0.5V Bulk Input Fully-Differential Operational Transconductance Amplifier," in *Proc. European Solid-State Circuits Conference (ESSCIRC)*, pp. 147–150, Sept. 2004.

[5.4] S. Chatterjee, Y. Tsividis and P. Kinget, "A 0.5V Filter with PLL-Based Tuning in 0.18µm CMOS," *IEEE International Solid-State Circuits Conference (ISSCC), Digest of Technical Papers*, pp. 506–507, Feb. 2005.

[5.5] J. Goes, N. Paulino, H. Pinto, R. Monteiro, Bruno Vaz and A. S. Garção, "Low-Power Low-Voltage CMOS A/D Sigma-Delta Modulator for Bio-Potential Signals Driven by a Single-Phase Schemee," *IEEE Transactions on Circuits and Systems-I, Regular Papers*, vol. 52, no. 12, pp. 2959–2604, Dec. 2005.

[5.6] M. Keskin, U. K. Moon and G. C. Temes, "A 1-V 10-MHz Clock Rate 13-bit CMOS ΔΣ Modulator using Unity-Gain-Reset Opamps," *IEEE Journal Solid-State Circuits (JSSC)*, vol. 37, no. 7, pp. 817–824, July 2002.

[5.7] G. C. Ahn, D.-Y. Chang, M. E. Brown, N. Ozaki, H. Youra, K. Yamamura, K. Hamashita, K. Takasuka, G. C. Temes and U.-K. Moon, "A 0.6V 82dB ΔΣ Audio ADC using Switched-RC Integrators," *IEEE International Solid-State Circuits Conference (ISSCC), Digest of Technical Papers*, pp. 166–167, Feb. 2005.

[5.8] S. Karthikeyan, S. Mortezapour, A. Tammineedi and E. K. F. Lee, "Low-Voltage Analog Circuit Design based on Biased Inverting Opamp Configuration," *IEEE Transactions on Circuits and Systems-II (TCAS-II)*, vol. 47, no. 3, pp. 176–184, Mar. 2000.

[5.9] T. Tille, J. Sauerbrey and D.-S. Landsiedel, "A 1.8-V MOSFET-Only ΣΔ Modulator using Substrate Biased Depletion-Mode MOS Capacitors in Series Compensation," *IEEE Journal of Solid-State Circuits (JSSC)*, pp. 1041–1047, July 2001.

[5.10] S. Pavan and Y. Tsividis, *High Frequency Continuous Time Filters in Digital CMOS Processes*, Kluwer Academic Publishers, 2000.

[5.11] P.-I. Mak, S.-P. U and R. P. Martins, "A 1-V Transient-Free and DC-Offset-Canceled PGA with a 17.1-MHz Constant Bandwidth over 52-dB Control Range in 0.35-µm CMOS," in *Proc. IEEE Custom Integrated Circuits Conference (CICC)*, pp. 649–652, Sept. 2005.

[5.12] P.-I. Mak, S.-P. U and R. P. Martins, "On the Design of Low-Voltage, Transient-Free and Constant-Bandwidth Programmable-Gain Amplifier with built-in DC-Offset Cancelers," submitted for journal publication.

[5.13] Jussila, J. Ryynänen, K. Kivekäs, L. Sumanen, A. Pärssinen and K. Halonen, "A 22mA 3.7dB NF Direct Conversion Receiver for 3G WCDMA," *IEEE International Solid-State Circuits Conference (ISSCC), Digest of Technical Papers*, pp. 284–285, Feb. 2001.

[5.14] Z. Xu, et al., "A Compact Dual-Band Direct-Conversion CMOS Transceiver for 802.11a/b/g WLAN," *IEEE International Solid-State Circuits Conference (ISSCC), Digest of Technical Papers*, pp. 98–99, Feb. 2005.

[5.15] R. Baker, H. W. Li and D. E. Boyce, *CMOS Circuit Design, Layout and Simulation*, IEEE Press, pp. 528–529, 1998.

[5.16] M. Dessouky and A. Kaiser, "Very Low-Voltage Digital-Audio Modulator with 88-dB Dynamic Range using Local Switch Bootstrapping," *IEEE Journal of Solid-State Circuits (JSSC)*, vol. 36, no. 3, pp. 349–355, Mar. 2001.

[5.17] A. S. Sedra and K.C. Smith, *Microelectronic Circuits*, 4th edition, Oxford University Press, 1998.

[5.18] J. A. Weldon, R. S. Narayanaswami, J. C. Rudell, L. Lin, M. Otsuka, S. Dedieu, L. Tee, K.-C. Tsai, C.-W. Lee and P. R. Gray, "A 1.75-GHz Highly Integrated Narrow-Band CMOS Transmitter with Harmonic-Rejection Mixers," *IEEE Journal of Solid-State Circuits (JSSC)*, vol. 36, pp. 2003–2015, Dec. 2001.

[5.19] P.-I. Mak, S.-P. U and R. P. Martins, "A Front-to-Back-End Modeling of I/Q Mismatch Effects in a Complex-IF Receiver for Image-Rejection Enhancement," in *Proc. IEEE International Conference on Electronics, Circuits and Systems (ICECS)*, pp. 631–634, Dec. 2003.

[5.20] K. Su, *Analog Filters*, 2nd edition, pp. 230–231, Kluwer Academic Publishers, 2002.

[5.21] C.-C. Hsu and J.-T. Wu, "A Highly Linear 125-MHz CMOS Switched-Resistor Progammable Gain Amplifier," *IEEE Journal of Solid-State Circuits (JSSC)*, vol. 38, no. 10, pp. 1663–1670, Oct. 2003.

[5.22] L. Breems, E. J. Zwan and J. H. Huijsing, "A 1.8-mW CMOS ΣΔ Modulator with Integrated Mixer for A/D Cconversion of IF Signals," *IEEE Journal of Solid-State Circuits (JSSC)*, vol. 35, no. 4, pp. 468–475, Apr. 2000.

[5.23] R. Harjani, J. Kim and J. Harvery, "DC-Coupled IF Stage Design for a 900-MHz ISM Receiver," *IEEE Journal of Solid-State Circuits (JSSC)*, vol. 38, no. 1, pp. 126–134, Jan. 2003.

[5.24] H. Darabi, J. Chiu, S. Khorram, H. J. Kim, Z. Zhou, H.-M. Chien, B. Ibrahim, E. Geronaga, L. H. Tran and A. Rofougaran, "A Dual-Mode 802.11b/Bluetooth Radio in 0.35-µm CMOS," *IEEE Journal of Solid-State Circuits (JSSC)*, vol. 40, no. 3, pp. 698–706, Mar. 2005.

Chapter 6

AN EXPERIMENTAL 1-V SIP RECEIVER ANALOG-BASEBAND IC FOR IEEE 802.11A/B/G WLAN

1. INTRODUCTION

This chapter reports the implementation details and experimental results of a SiP receiver analog-baseband (BB) IC for IEEE 802.11a/b/g WLAN. A new test strategy is introduced, which allows both functional-block and full-system measurements. Fabricated in a conventional 3.3-V 0.35-µm CMOS process without resorting to any specialized technology option, the verified techniques are directly migratable and are expected to yield higher performances in the forthcoming sub-1 V processes, of which the threshold voltages, forecasted by the *International Technology Roadmap of Semiconductors* (ITRS) [6.1], will be within 0.2–0.3 V in later technologies for high-tier wireless applications.

2. RECEIVER ARCHITECTURE

Figure 6-1 depicts the proposed flexible-IF receiver, where the implemented analog-BB IC is bracketed. The RF front-end consists of a dual-band LNA and a pair of dual-band mixers operated in quadrature. By them, the radio channels are downconverted to either a low IF or dc. The foremost filtering are performed by dual center-frequency-tunable (i.e., +IF, –IF or 0) second-order Butterworth biquad filters that have HPFs embedded. They in low-IF mode, together with the preselect filter, prevent the residual in-band channels and out-of-band white noise, in the subsequent IF-to-BB downconversion, from folding back within the signal band. The double-quadrature downconverter (DQDC) is made up of a series-switching (SS) mixer-quad. It is driven by a multiphase clock generator (CLKGEN) to realize a wideband mismatch-insensitive I/Q demodulation. In order to complete the second step of channel selection, the DQDC is designed to have a sideband selectivity (i.e., switch the phase (0°, 180°) of its I/Q-coupled paths). The downconversion is performed before filtering and amplification, such that the mode reconfiguration from LIF to ZIF are, simply, halving the BW of the preselect filter and disabling the DQDC and CLKGEN. For power and area savings, a 3-stage 17-MHz-constant-BW PGA is adopted, instead of a wideband PGA, which is conventionally essential for maintaining the selectivity constant against gain. In this work, a twofold relaxation of the Butterworth LPF's order from fifth to third is attained.

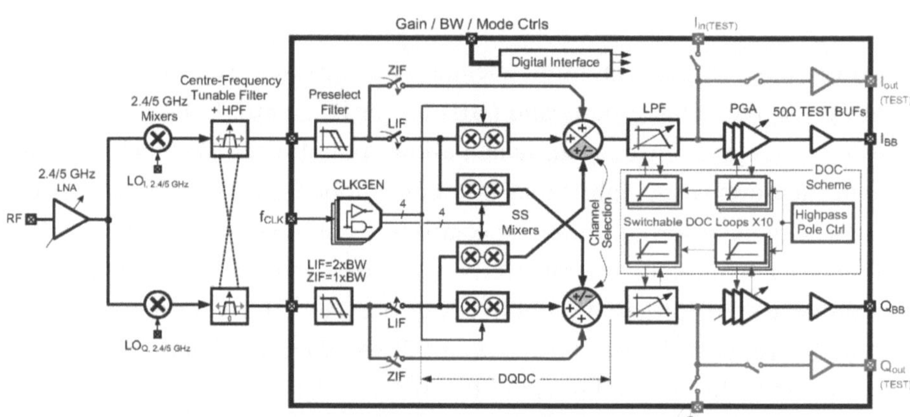

Figure 6-1. Detailed architecture of the proposed flexible-IF receiver

A dc-offset-canceler (DOC) scheme using an inside-OpAmp technique is proposed. A switchable DOC loop is embedded inside each LPF's and PGA's OpAmp, by which the differential signals are locally balanced and the composite highpass pole is agilely switchable for tracking the dc-offset transient. In preamble reception, all DOCs are switched off such that the receiver settling time is only governed by the ac-coupler (not shown in the figure) that eventually interfaces the PGA to the A/D converter (9- to 10-bit resolution [6.2]). The ac-coupler is designed to have an extended lower –3-dB point of 1 MHz to receive just the pilot tones. In ZIF mode, all DOCs are switched on but maintaining a composite lower –3-dB point less than 10 kHz. This low-frequency value ensures a low intersymbol interference in processing CCK channels. The gain, BW and mode are all controlled digitally. A built-in setup and additional 50-Ω test buffers enable both full-chip and functional-block measurements.

3. SIMULATION METHODOLOGY

The scope of a flexible-IF, low-voltage, low-power receiver analog baseband requires innovative design and simulation permutations. The simulation methodology with a mixture of available design tools is summarized in Figure 6-2.

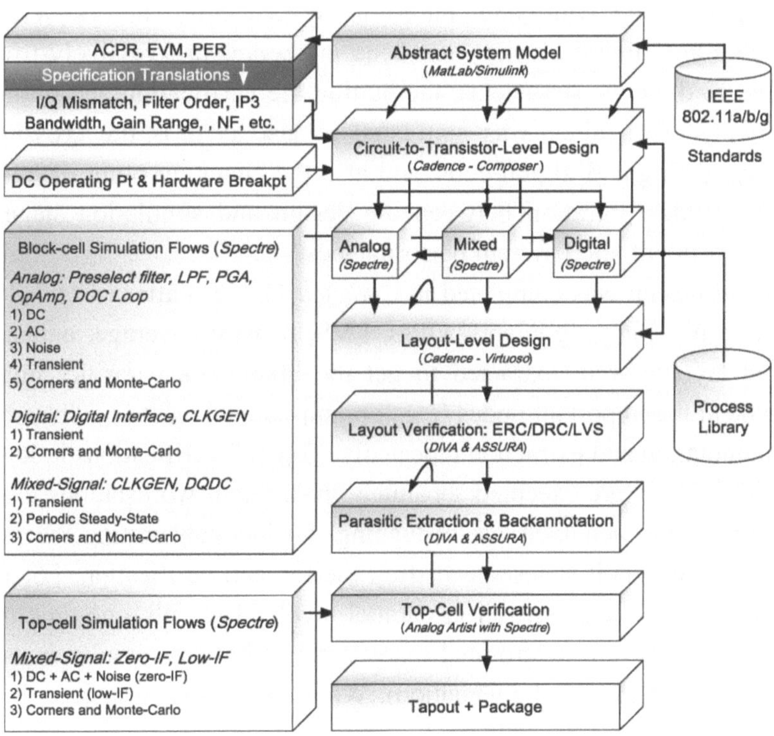

Figure 6-2. Simulation methodology

An abstract-system model is built by using *MatLab/Simulink* and the provided toolboxes for functionality verifications and virtual measurements of adjacent-channel power ratio (ACPR), error-vector magnitude (EVM) and packet-error rate (PER). Results from the models were translated to circuit-level specifications such as I/Q mismatch, filter order, third-order intercept point (IP3) and noise figure (NF), etc. Circuit-to-transistor-level design was carried in *Cadence Composer*. The dc-operating points (e.g., input and output common modes) should be set before circuit architectures were chosen, on the ground that differential pair and analog switches only work on the dc level close to one of the supply rails. Hardware breakpoints were assigned for block measurements in case of partial malfunction and for block net-response measurements. All analog cells were full-custom designed, whereas the digital ones were based on standard cells with a scaled-up physical size to compensate the increased delay of low-voltage implementation (gate delay $\propto$ 1/supply). Constraint-driven optimization reduced the number of iterations in the design of paramount analog cells such as OpAmp gain, bandwidth and phase-margin trade-offs. The first verifications of analog and digital blocks were done in *Spectre* progressively from static (dc, ac) to dynamic (transient), facilitating the verification and optimization effectively. The mixer was simulated at last as it is the crossing point between analog and digital cells, and also the changing point of low-IF and zero-IF modes. Iteration between the design and simulation stages continued, dependent upon system performance.

Layout design was completed in *Cadence Virtuoso* after careful floorplanning and pin-out assignments. DRC, LVS, density coverage, antenna-effect cleared layouts were extracted to get the ubiquitous parasitic. Imbalanced parasitic at the important nodes (e.g., op-amp inputs) were equalized by doing back annotation and extraction repeatedly. Top-cell verifications in low-IF and zero-IF modes were executed separately in different workstations to save the simulation time and facilitate debugging. Deliberately setting the input and changing the circuit states speeded up the top-cell verification. For instance, dc, ac and noise analysis run in parallel involved in only one-time netlisting and dc analysis, but determined many static behaviors simultaneously (e.g., bandwidth and stopband attenuation). Whereas a liberal transient simulation, with an input level that could get the largest output swing, and with the largest gain step applied in between, exhibits the worst measured settling time and overshoot. Monte-Carlo simulations (dc, ac and selected transient) were executed iteratively before tape-out.

4. CIRCUIT IMPLEMENTATION

4.1 Preselect filter and DQDC and CLKGEN

Figure 6-3 shows the I-channel (Q-channel is identical) dual-mode preselect filter and DQDC (bypassable at ZIF mode). The front-end resistor-capacitor ($R_{PF}C_{PF}$) network offers an additional preselect lowpass filtering [6.3] with two possible BWs to band-limit the input signal and offer a linear voltage-to-current conversion for the DQDC. The value of R_{PF} determines the noise figure (NF) of the entire analog-BB IC. With R_{PF} = 2.5 kΩ, the desired specifications (<30 dB NF) are safely met. The IF channel selection is executed in the background, inside the CLKGEN, by switching phases SW_Q and SW_Q'. The DQDC arranged in a double-balanced structure cancels the unwanted I/Q demodulating carrier at the output. The swapper and mixer-switch S_M, driven by the depicted clock phases, cogenerate the two quasi-I/Q pulse trains, i.e., I = [...1, 0, –1, 0...] and Q = [...0, 1, 0, –1...], with a frequency that is a quarter of the reference CLK. As mentioned in Chapter 5, mismatch in the CLK duty cycle results in only quadrature amplitude mismatch but not phase. Moreover, as long as the edges are triggered within the tolerable timing margins (i.e., $T_{AE1} \leq 25/20$ ns and $T_{AE2} \leq 50/40$ ns for 10/12.5-MHz IF), the variation of the transitions do not affect the quadrature-phase accuracy of the mixer. The conversion gain (CG) is 0.45 (25% duty cycle), equivalent to a 7-dB loss in noise figure (NF), which it is practically affordable by a 30-dB-gain 5-dB-NF RF front-end [6.4]. The overall receiver NF is expected to be around 8.5 dB.

Comparing with the technique reported in [6.5], the proposed I/Q generation can tolerate similar amount of I/Q mismatches but featuring a higher CG while requiring a lower CLK (in CG is 0.24 and CLK has to be 8× the IF).

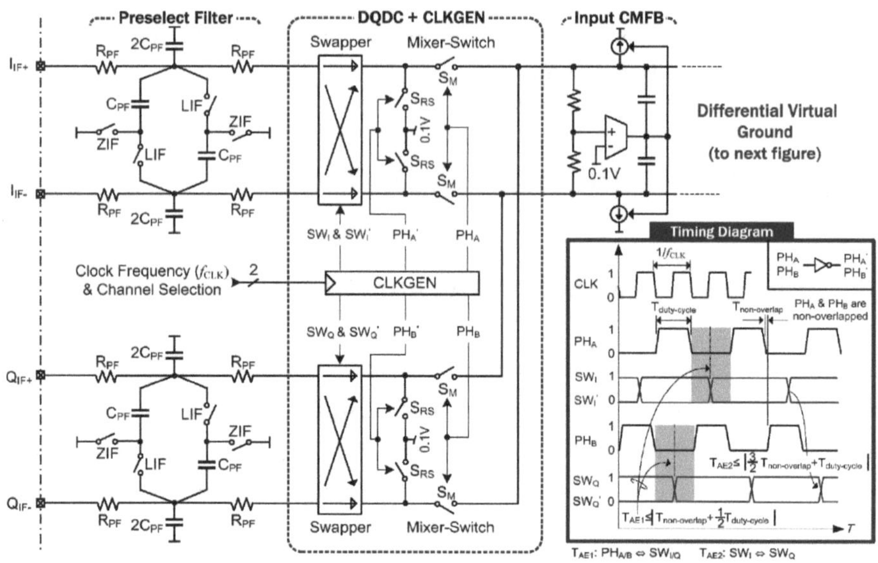

Figure 6-3. I-channel dual-mode preselect filter and double-quadrature downconverter (DQDC)

In order to achieve a very linear mixing operation, the on-resistance r_{on} of the swapper and S_M is designed to be much lower than that of R_{PF}, and the common-mode voltage of the differential virtual ground $V_{cm,in}$ is biased at its lowest possible value (i.e., 0.1 V) by using an input common-mode feedback circuit (CMFB). In our case, r_{on} can be as large as 213 Ω (8.5% of R_{PF}) for a third-harmonic distortion (HD3) of –80 dB.

Other advantages of the SS mixing are also worth to mention: (1) The swapper, operating at an identical frequency with the IF, is activated only when S_M is in open state, avoiding any charge injection from the swapper and self-mixing in overall. Of course, S_M itself induces charge injection, but it is out of the signal band (i.e., 2× the IF); (2) The reset-switch S_{RS} reduces the conversion loss and memory effect during swapping of the differential branches; (3) The swapping-induced charge coupling back to the RF front-end is suppressed by the preselect filter.

4.2 Channel-selection LPF, PGA and DOC scheme

The filtering and amplification are codesigned for not only leveraging the linearity and noise, but also minimizing the power and area. As shown in

Figure 6-4, a third-order Butterworth active-resistance-capacitance (*RC*) LPF (1 uniquad + 1 biquad with Q = 1.3065) incorporates a 3-stage 17-MHz-constant-BW PGA to attain a constant selectivity better than a typical fifth-order LPF, as well as offering a controllable gain ranging from –2 dB to 50 dB with 2-dB per step. Two coarse-stage (6 dB/step) followed by a fine-stage (2 dB/step) gain controls were utilized in the PGA for achieving low global transients. Through iterative simulations and with a positive zero (R_{ff} and C_z) added in the PGA's 3rd-stage, the optimized (through simulation) group-delay peaking at the band edge is 14.8 ns. The resistor R_{BW} is a resistor array for tuning the BW digitally. The switched-current-resistor (SCR) technique (Chapter 5) has been employed for a transient-free constant-BW gain adjustment.

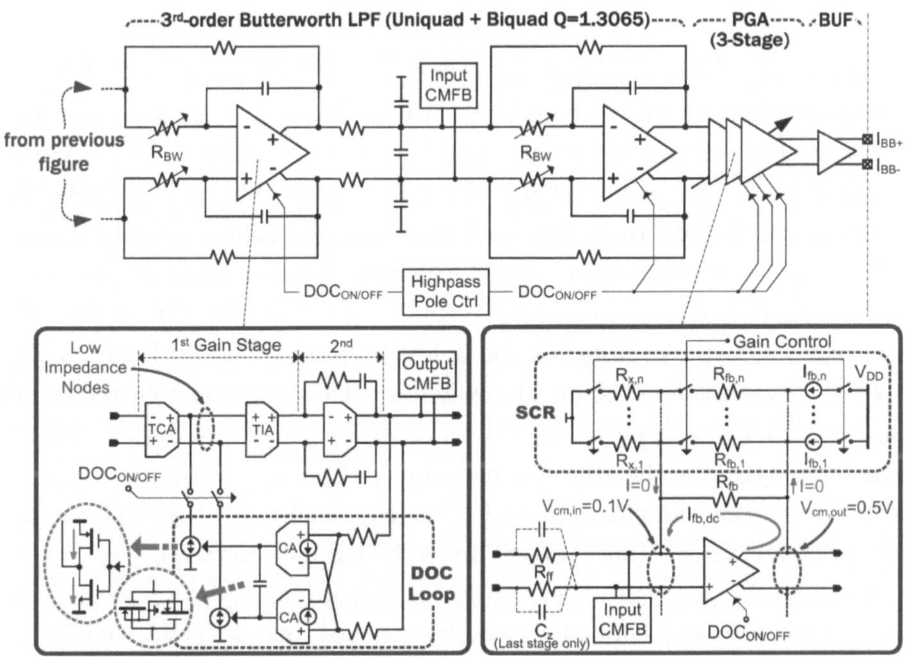

Figure 6-4. I-channel channel-selection LPF and PGA (lower-left: OpAmp with built-in DOC loop, lower-right: one SCR PGA stage)

To improve the linearity and prevent the large BB gain from saturating the system, each LPF and PGA's stage has an individual DOC loop. Unlike the conventional servo loop that typically closes the feedback with multiple forward stages, an individual DOC loop alleviates the stability consideration.

Taking the advantageous properties of negative feedback in bandwidth extension [6.6], each OpAmp has a built-in switchable servo loop (Figure 6-4 lower left), such that in closed loop use inside the LPF and PGA, the frequency of the highpass pole will be lowered by an amount of loop gain. The transistor-level implementations of the OpAmp and the DOC loop were summarized in Chapter 5. All LPF and PGA stages employ the identical OpAmp and DOC-loop structure, with the foremost front stages optimized for low noise, while the backmost are tapered in gain not to compromise the linearity.

4.3 Design of I/O circuitry

Chip-to-chip bonding requires the concern of I/Os since each chip has ESD protection and may use different voltage levels in their pad rings. For instance, the inductance of the bondwire and the parasitic capacitance associated with the pads that will limit the bandwidth, and the potential difference of the pad rings that will limit both the common-mode voltage and signal swing. In the analog BB chip implementation, standard pads were employed since only low-frequency signals were transferred in and out the BB chip. The bypass settings in the 3.3-V I/Os are shown in Figure 6-5(a) and (b), respectively. Placing the switches after the resistor ensured high linearity and low-voltage workability. Two single-ended test buffers (BUFs) were employed for the differential outputs (Figure 6-5(b)). Inserting an inverter in the current mirror, the BUF can be switched off (on) with its gate voltage equal to 3.3 V (bias voltage driven by current source I_b).

Although the pad-ring upper and lower voltage limits are set to 0 V and 3.3 V, respectively, simulation results showed that a THD <0.03% is achieved with a 1.2-Vpp input at 0.1-V dc level (Figure 6-6). This implies that the signal swing can go beyond the pad-ring limits by a voltage level close to the forward-bias voltage of the protection diodes. Thus, unmatched pad-rings may not pose a significant limitation to chip-to-chip interconnect.

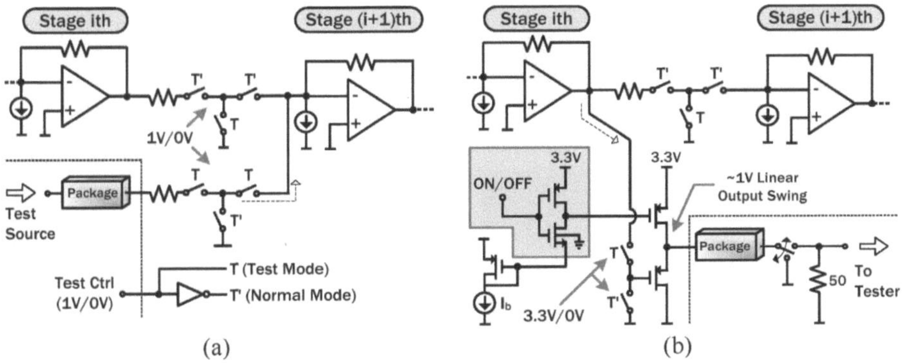

Figure 6-5. I/O setup for (a) test-source injection and (b) measurement

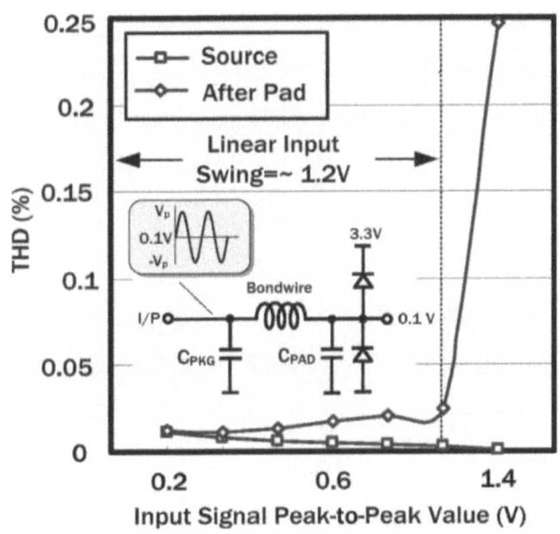

Figure 6-6. Package linearity to input swing at 0.1-V dc level

5. SIMULATION RESULTS

Figure 6-7(a)–(d) summarize post-layout simulation results. The output white-noise density at maximum gain is ~10 μV/√Hz, while the 1/*f* noise is suppressed by more than 32 dB due to the highpass characteristic (Figure 6-7(a)). Because of the use of a constant-feedback-factor PGA, the lower and upper –3-dB points remain practically constant at all gain levels (Figure 6-7(b)). The robustness of the chip is demonstrated in a 100-time AC Monte-Carlo simulation (Figure 6-7(c)), no ill yield is observed while process variations and mismatch are taken into account concurrently. The worst spurious-free dynamic range (SFDR) is 46 dB with 0.98-V_{pp} output swing (Figure 6-7(d)).

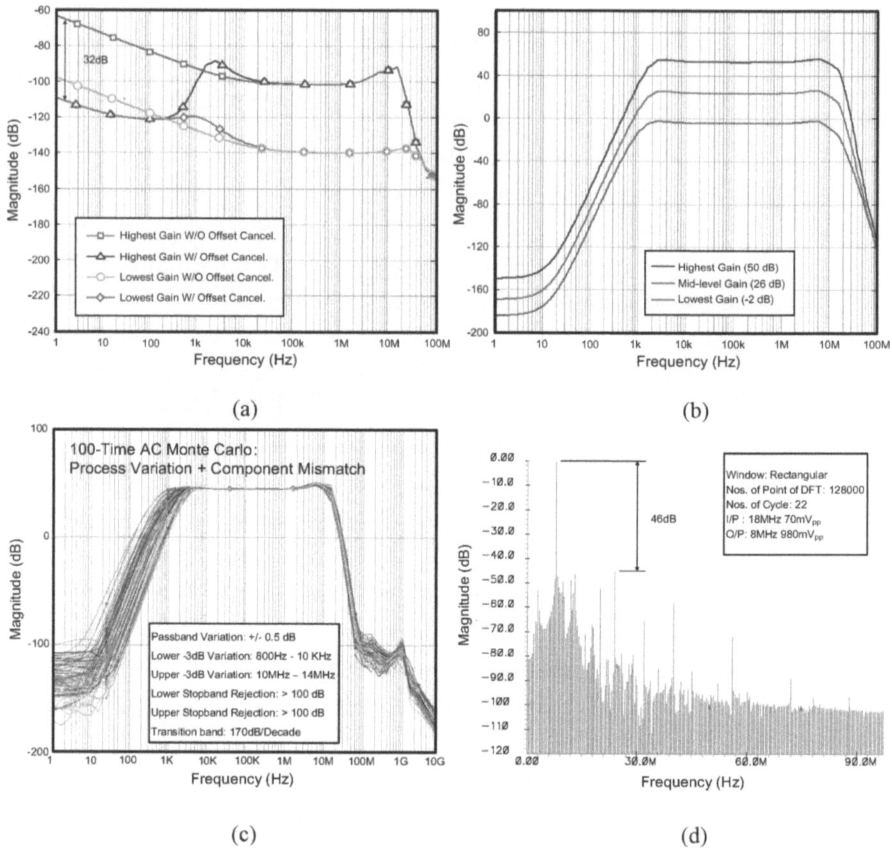

Figure 6-7. Post-layout simulation results: (a) output noise at highest and lowest gain, with/without offset cancellation (ZIF) (b) AC response (ZIF) (c) Monte-Carlo AC responses in ZIF mode (d) FFT response in LIF mode

6. SILICON IMPLEMENTATION AND TEST STRATEGY

The fully differential analog-BB IC [6.7][6.8] is fabricated in a 3.3-V 0.35-μm double-poly five-layer metal CMOS technology and is housed in a ceramic quad flat pack (CQFP) [6.9]. The chip micrograph is shown in Figure 6-8. The design uses separated V_{DD} supply pins but shared common ground with on-chip decoupling (MOS capacitors) for analog and digital parts to minimize the inductance in the current return path for signal transfer. A strict symmetry and matching in sensitive circuits like the DQDC has been maintained together with a clean signal and substrate environment achieved by ample substrate contacts. In total ten DOC loops were integrated inside each LPF and PGA's stage. Core area is 3 mm^2, with 0.02 mm^2 per DOC loop. The test buffers (BUFs) are 3.3-V source followers.

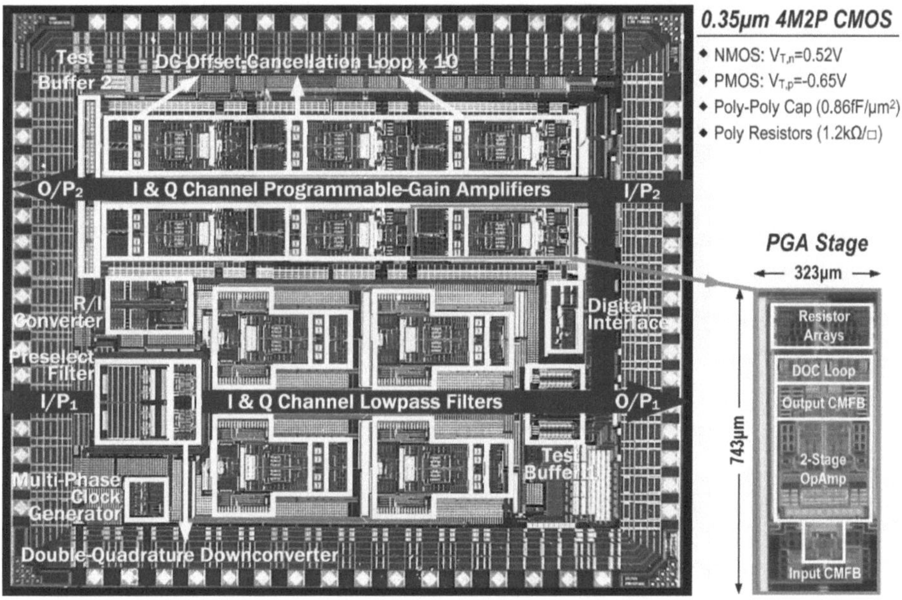

Figure 6-8. Chip micrograph

According to the design methodology [6.10], block net-responses were intended to be measured individually. The evaluation board is full-custom [6.11] EMI-aware designed, which include passive inductance-capacitance (*LC*) filtering of CM voltages, individually tunable bias currents (SMD LM334 [6.12] and simple tunable resistors are both included for generating 100-μA references) and differential and single-ended 50-Ω testing ports, as depicted in Figure 6-9 and Figure 6-10. Star-center routing is set, with the closest surroundings occupied with decoupling capacitors. Clock signal is injected straightly to the chip. The outmost parts are occupied by the voltage and current sources. The measurement equipments used are listed in Table 6-1 and the experimental setup is depicted in Figure 6-11, in which a grounded diecast for EMI-shielding is custom made to improve the quality of measurements.

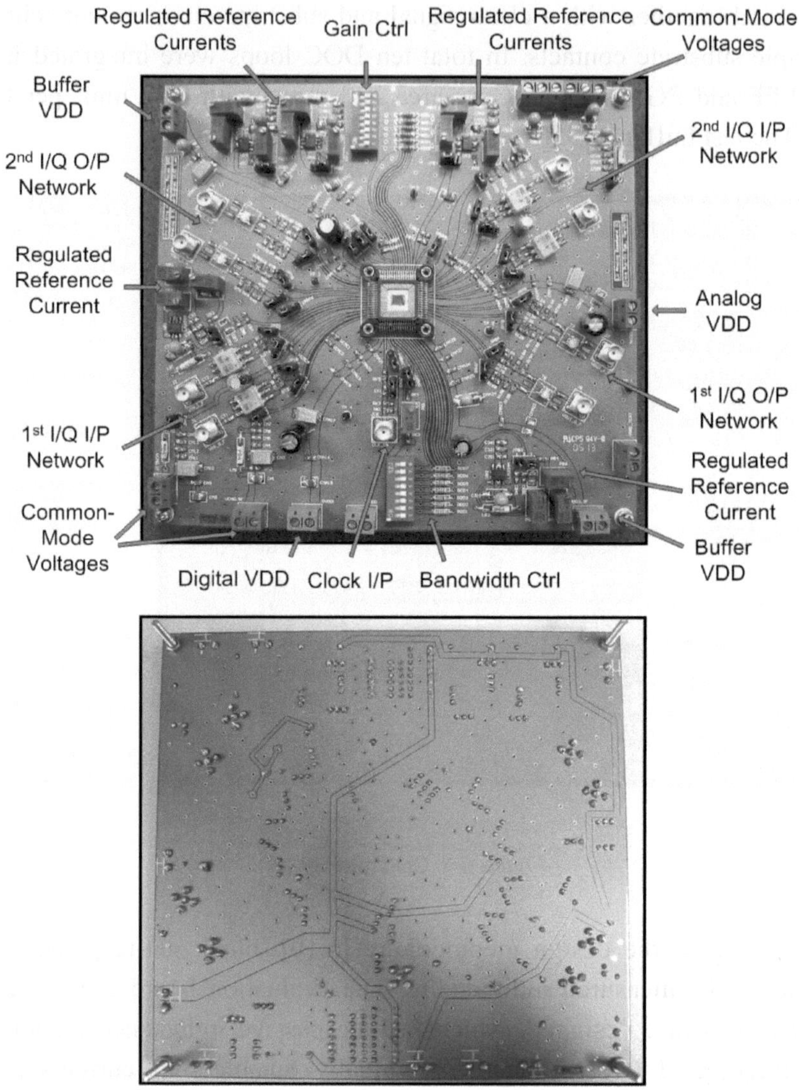

Figure 6-9. Evaluation board (upper: top view and lower: bottom view)

Table 6-1. Testing equipments

Rohde & Schwarz FSU8 Spectrum Analyzer (20 Hz – 8 GHz)	Agilent 33250A 80 MHz Function/Arbitrary Waveform Generator
Rohde & Schwarz FSU-9 Tracking Generator (100 kHz – 3.6 GHz)	Agilent E4424B ESG-AP Series Signal Generator (250kHz – 2GHz)
Rohde & Schwarz FS-K30 Noise and Gain Measurement	Agilent E4436B ESG-DP Series Signal Generator (250kHz – 3GHz)
HP 35670A Dynamic Signal Analyzer (0 Hz – 51.2 kHz for AC)	Agilent E4430BK-UN8 Real-Time I/Q Baseband Generator 1M RAM
Mini-Circuits ZFSC-2-6 2 Way-0° Power Splitter/Combiner (2kHz – 60MHz)	Agilent Infiniium 54832D Oscilloscope 1 GHz, 4 GSa/s
Mini-Circuits Transformers – TCM1-1 (1.5 MHz -500 MHz) TT1-6-KK81 (0.004 MHz -300 MHz)	Agilent Probes: 1161A Passive, 1159A 1 GHz Differential, 1157A 2.5G Hz Active, E2621A and E2622A Termination Adaptors
Mini-Circuits BNC/SMA Passive Filters – BBP10.7, BBLP117, VLF80	Agilent 34420A 7.5 Digit Nano Volt/Micro Ohm Meter
Yamaichi IC149-064-008-S5 QFP Production-Use Socket	Agilent - E3631A DC Power Supply (0-6V/5A, 0- ±25V/1A)
Fluke 179 True RMS Multimeter	LitePoint IQnxn Vector Signal Generator-Analyzer
National Semiconductor LM334 Current Regulator	Hammond IP54 RFI/EMI Shielded Diecast Box

For testability, a specific floor plan and on/off-chip bypass setup are developed for full-chip and block net-response measurements that are independent to the characteristics of the I/O networks and test BUFs. As shown in Figure 6-12, the floor plan includes on/off-chip bypassed switches and two matched 50-Ω BUFs. As listed in Table 6-2, the proposed technique calls for seven measurements – cases {1}–{3} allows the examination of the net responses of the I/O networks whereas rest cases are to determine the responses of: {1} the preselect filter plus DQDC and LPF; {2} the preselect filter plus LPF; {3} only the PGA and {4} the overall chip. After simple multiplications and divisions, the net responses of the functional blocks: {10}–{12}, and the overall responses: {13}–{14}, can be found.

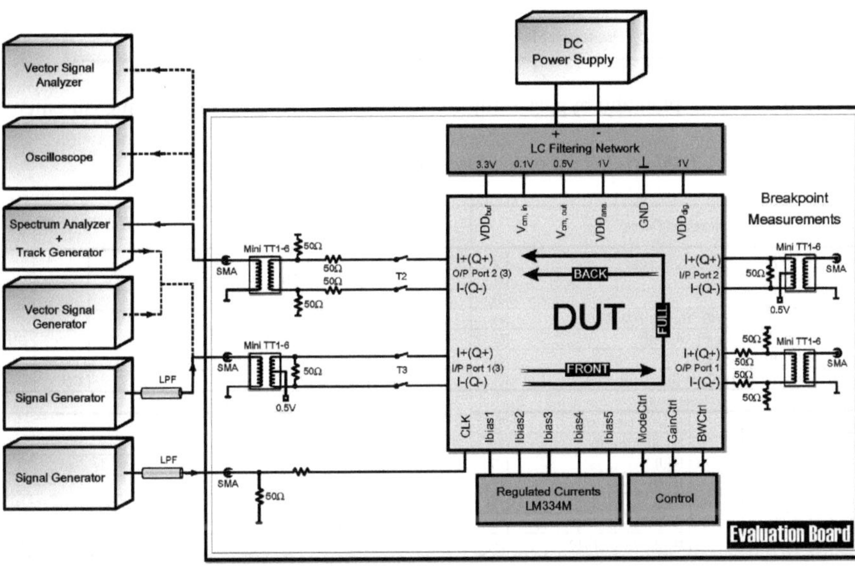

Figure 6-10. Evaluation board floor plan

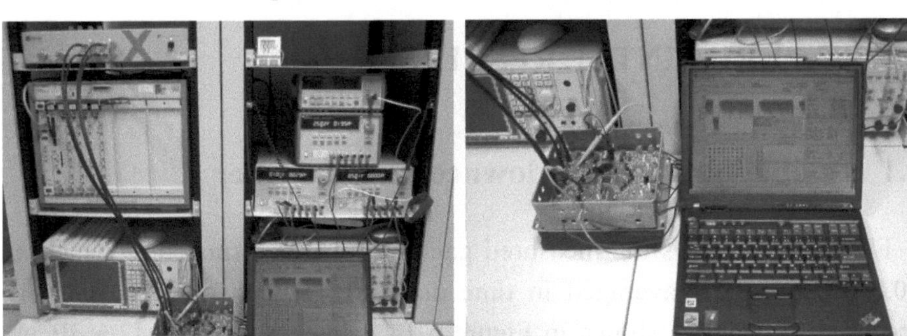

Figure 6-11. Experimental setup (constellation diagram and EVM test)

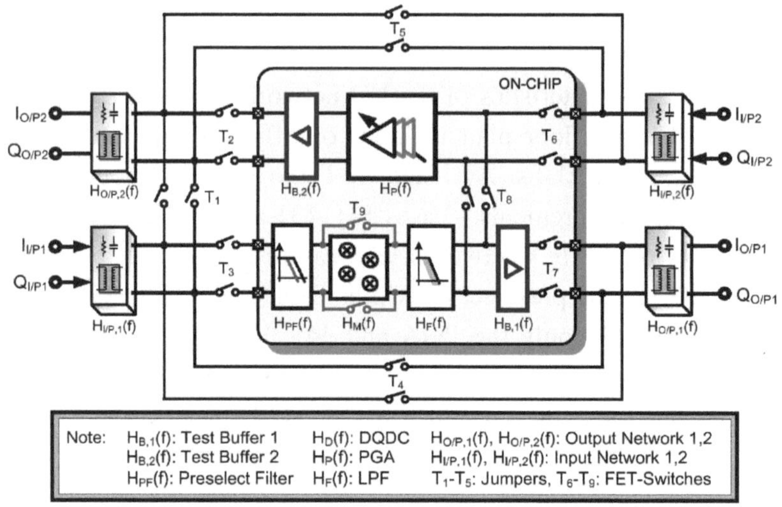

Figure 6-12. Floor plan and bypass setup for functional-block net-response measurements

Table 6-2. Test procedures corresponding to Figure 6-12

Measured Cases		Off-Chip Switches					On-Chip Switches			
		T_1	T_2	T_3	T_4	T_5	T_6	T_7	T_8	T_9
$H_{I/P,1}(f) \cdot H_{O/P,2}(f)$	{1}	ON								
$H_{I/P,1}(f) \cdot H_{O/P,1}(f)$	{2}				ON					
$H_{I/P,2}(f) \cdot H_{O/P,2}(f)$	{3}					ON				
$H_{I/P,1}(f) \cdot H_{PF}(f) \cdot H_M(f) \cdot H_F(f) \cdot H_{B,1}(f) \cdot H_{O/P,1}(f)$	{4}			ON				ON		
$H_{I/P,1}(f) \cdot H_{PF}(f) \cdot H_F(f) \cdot H_{B,1}(f) \cdot H_{O/P,1}(f)$	{5}			ON				ON		ON
$H_{I/P,2}(f) \cdot H_P(f) \cdot H_{B,2}(f) \cdot H_{O/P,2}(f)$	{6}		ON				ON			
$H_{I/P,1}(f) \cdot H_{PF}(f) \cdot H_D(f) \cdot H_F(f) \cdot H_P(f) \cdot H_{B,2}(f) \cdot H_{O/P,2}(f)$	{7}		ON	ON					ON	

Operations					Net Responses	
{2} / {1}	→ get	$H_{O/P,1}(f) / H_{O/P,2}(f)$	{8}	→ get	Output-Network Ratio	
{3} / {1}		$H_{I/P,2}(f) / H_{I/P,1}(f)$	{9}		Input-Network Ratio	
{4} / {5}		$H_D(f)$	{10}		DQDC	
{9} · {7} / ({6} · {10})		$H_{PF}(f) \cdot H_F(f)$	{11}		Preselect Filter + LPF	
{8} · {7} / {4}		$H_P(f)$	{12}		PGA	[Assume $H_{B,1}(f)=H_{B,2}(f)$]
{10} · {11} · {12}		$H_{PF}(f) \cdot H_D(f) \cdot H_F(f) \cdot H_P(f)$	{13}		Preselect Filter + DQDC + LPF + PGA	[LIF Modes]
{11} · {12}		$H_{PF}(f) \cdot H_F(f) \cdot H_P(f)$	{14}		Preselect Filter + LPF + PGA	[ZIF Mode]

7. EXPERIMENTAL RESULTS

7.1 Double-quadrature-downconversion filter

Figure 6-13 shows the measured gain and phase mismatches, which are <0.5 dB and 0.5° (averaged in time domain), respectively. The complex spectrum of |I+jQ| is shown in Figure 6-14, where an IRR of 42.6 dB was measured. On the other hand, shown in Figure 6-15 are the gain responses at low and high frequencies, the lower –3-dB frequency is switchable and the BW control of the stand-alone CSF shows a tunable range between 4.34 and 10.18 MHz. The in-band IIP2 and IIP3 with and without the DOC loops activated are shown in Figure 6-16(a) and (b), respectively. The result shows a significant improvement in IIP2 from 27.2 to 59 dBm. The IIP3 is improved as well since the third-harmonic modulates with the dc tone that is suppressed. The dc gain is 20 dB as designed (accurately determined by resistive ratio). The performance summary is listed in Table 6-3.

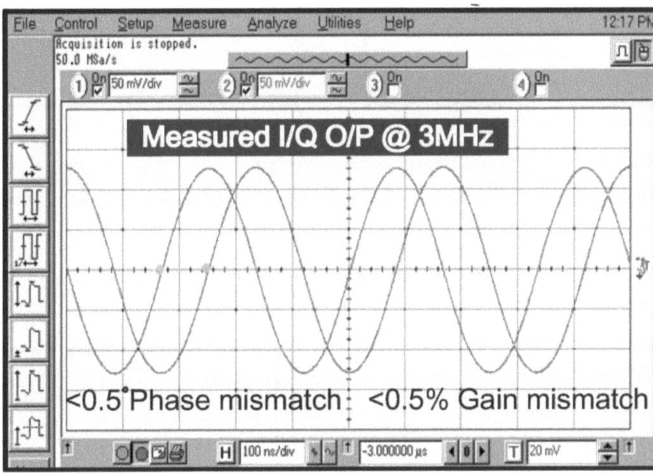

Figure 6-13. A snapshot of the oscilloscope's screen showing the I/Q output waveform (time averaged)

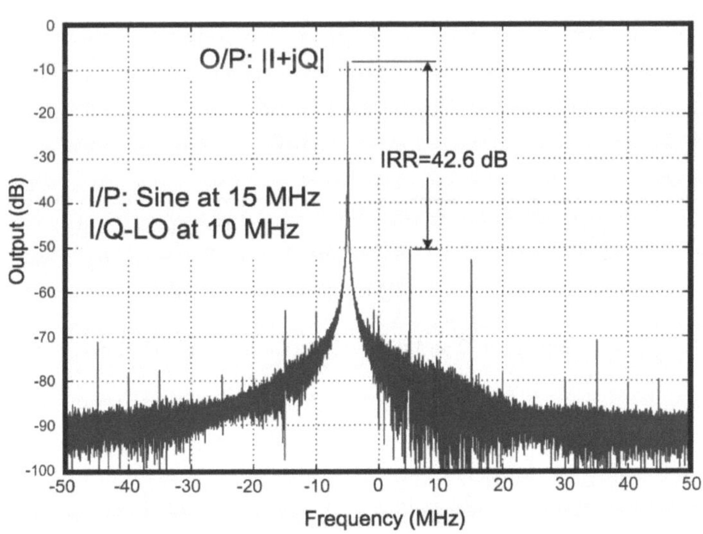

Figure 6-14. Spectrum of |I+jQ| with 15-MHz input downconverts to 5 MHz

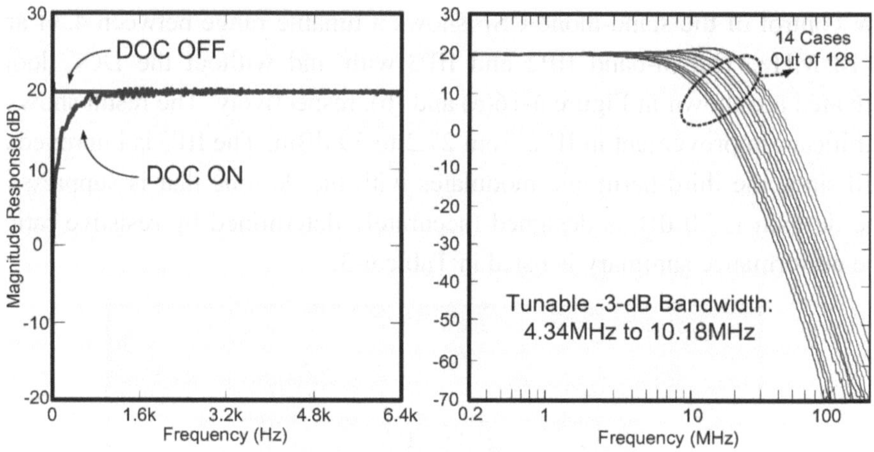

(a) (b)

Figure 6-15. (a) DOC ON/OFF magnitude response (b) Tunable BW of the CSF

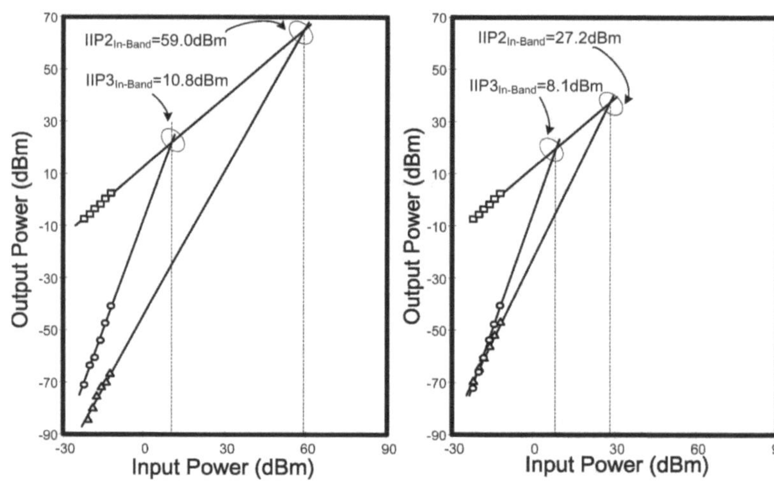

(a) (b)

Figure 6-16. Inband IIP2 and IIP3 with (a) and without (b) DOC loops

Table 6-3. DQDF measurement summary

Supported IFs		0/10/12.5 MHz
In-band IIP2/IIP3	With DOC Loops	+59.0/10.8 dBm
	Without DOC Loops	+27.2/ 8.1 dBm
Reference clock frequencies (4 × IF)		40/50 MHz
Output Noise Density (white)		57.1 nV/√ Hz
Filter Order		1st (single-pole) + 3rd (Butterworth)
IF Channel-Selection Transient		< 0.38 μs
CSF Tunable Bandwidth		4.34 to 10.18 MHz
Gain and Phase Mismatches		<0.5 dB, <0.5°
Current Consumption ‡		12.2 mA
Supply Voltage		1 V
Technology ($V_{th,n}$=0.52 V, $V_{th,p}$= –0.65 V)		0.35μm CMOS
Active Area of one channel /one S-OC		1.42 mm² / 0.02 mm²

‡: Exclude the test buffers.

7.2 SCR programmable-gain amplifier

The gain responses measured at 20/4/–22-dB gain levels are shown in Figure 6-17 (a) and (b) for low- and high-frequency portions, respectively. The means of the lower –3-dB points are 2.25 kHz (σ = 11.2%), and 17.1 MHz (σ = 8.3%) for the upper value. The BW variation is dominated by the gain steps ≤ –10 dB due to insufficient low output impedance in driving the small resistive loads. The spurious-free dynamic range (SFDR) is 56.2/39.7 dB with/without dc-offset cancellation activated (Figure 6-18(a) and (b)), verifying the effectiveness of the DOC loops. The designed gains versus the obtained, gain error, output offset voltage V_{os} are plotted together in Figure 6-19. Linear-controlled gains were achieved with a 0.013-dB gain-error variance but with a positive offset measured in all cases. The noncanceled V_{os}, increases accordingly with gain but fluctuates in between, which we believe it is due to random mismatch internally. The canceled V_{os}, that is practically nonzero, is close to 5.7 mV. The transient in gain change was measured with a 52-dB gain step applied (Figure 6-20(a)). The output signal reaches the desired gain level in 0.2 μs. In between, the highpass pole causes small transient and settles within 266 μs. Figure 6-20(b) shows the dynamic behavior of the DOC when it is switched. No noticeable transient happens at the start and stop slots, and the offset voltage was canceled within 305 μs when all DOCs are switched on simultaneously. The overshoot is within the output signal swing (i.e., no hard distortion due to clipping). It is noteworthy that the circuit is free from transient when all DOCs are switched off due to the removal of the highpass pole. The performance summary is given in Table 6-4.

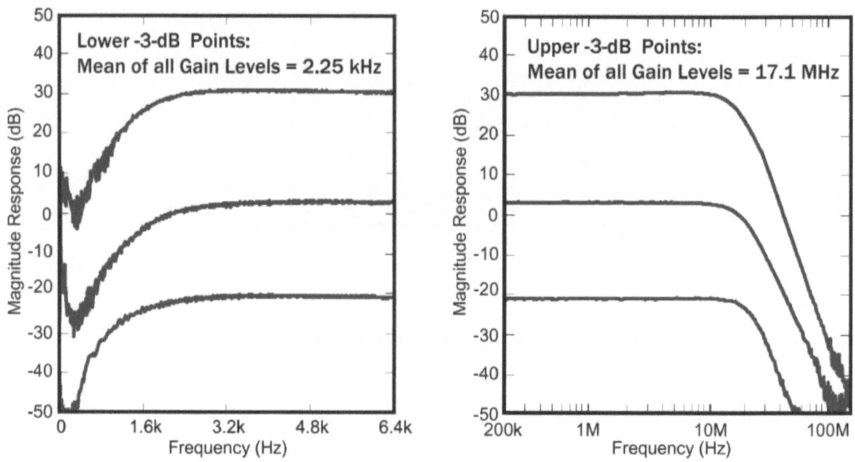

(a) (b)

Figure 6-17. Measured magnitude responses at 30/4/–22-dB gain levels: (a) low and (b) high frequency

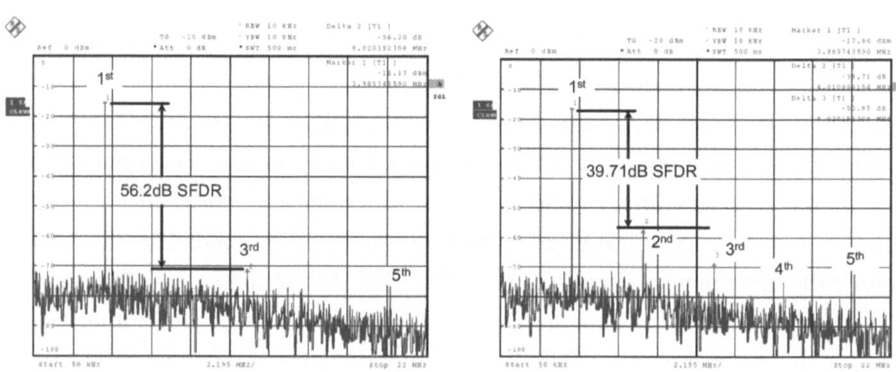

(a) (b)

Figure 6-18. SFDR with a 4-MHz single-tone input (a) with and (b) without dc-offset cancellation

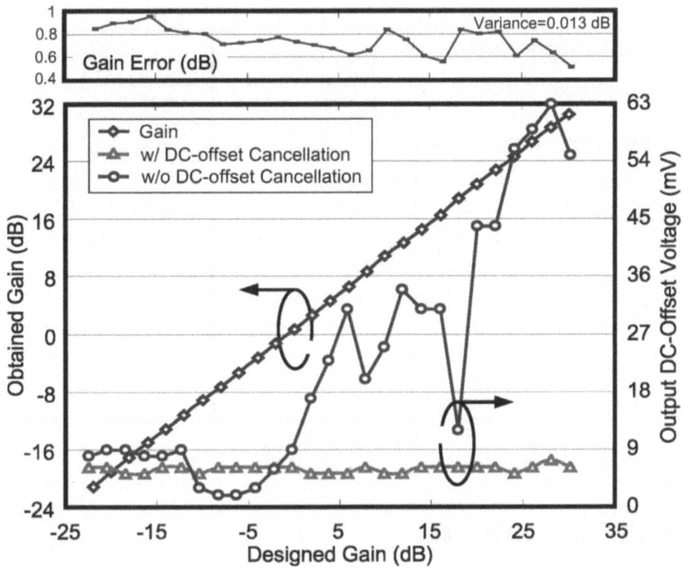

Figure 6-19. Designed gain versus output gain, gain error, output dc offset with/without dc-offset cancellation

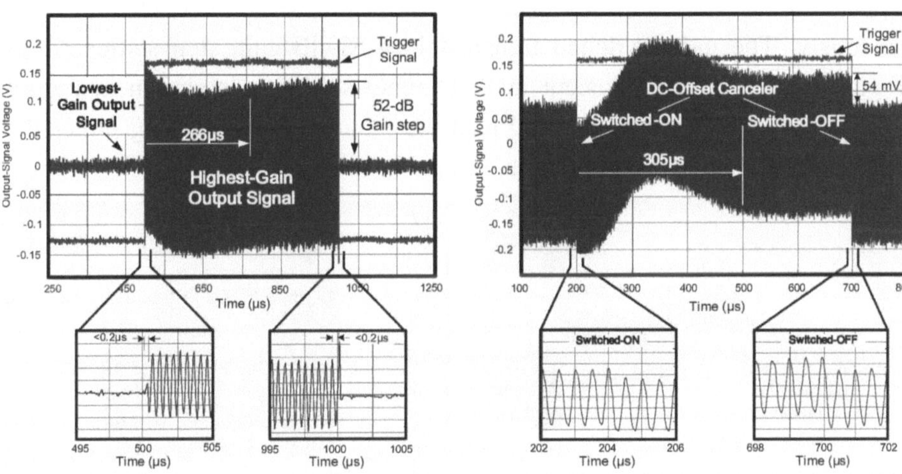

(a) (b)

Figure 6-20. PGA's dynamic performances: (a) DOC-loop switching (b) gain switching

Table 6-4. PGA measurement summary

Voltage Gain Range (2 dB/Step)		–22 ... +30 dB
Lower/Upper –3-dB Point	Mean of All Gains	2.25 kHz / 17.1 MHz
	Standard Deviation	11.2 % / 8.3 %
Output Noise Density (white)	At 30-dB Gain	273.1 nV/√ Hz
	At –22-dB Gain	20.5 nV/√ Hz
Transient Time tested by a 52-dB gain step		< 0.2 μs
In-Band IIP3		+ 8.4 dBm
SFDR (f_{in}=4 MHz, Gain Level=30 dB)		56.2 dB / 39.7 dB *Δ
Output Offset Voltage V_{os} †		5.7 mV / 23 mV *
Current Consumption †‡		7.4 mA
Supply Voltage		1 V
Technology ($V_{T,n}$: 0.52 V, $V_{T,p}$: –0.65 V)		0.35 μm CMOS
Active Area of one channel /one S-OC		0.72 / 0.02 mm²

†: Mean of all gains. ‡: Exclude the test buffers.
*: Without dc-offset cancellation. Δ: second harmonic dominant.

7.3 Analog-baseband IC

The dynamic performances were also fully characterized by in full-chip measurements. For the DOC loop (Figure 6-21(a)), no noticeable transient happens at the start/stop slot when all DOC loops are switched on simultaneously. The gain-switched transient in a 52-dB-gain step settles <1 μs (Figure 6-21(b)). The transient in IF channel selection is shown in Figure 6-22, where the *Q* output can lag/lead the *I* output for obtaining either the upper or lower sideband in 0.38 μs.

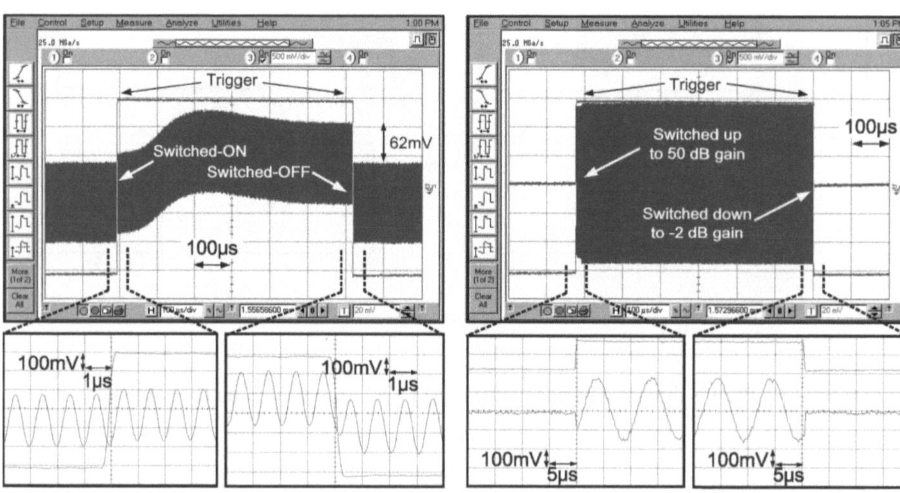

Figure 6-21. Analog-baseband chain's dynamic performances: (a) DOC-loop switching and (b) gain switching

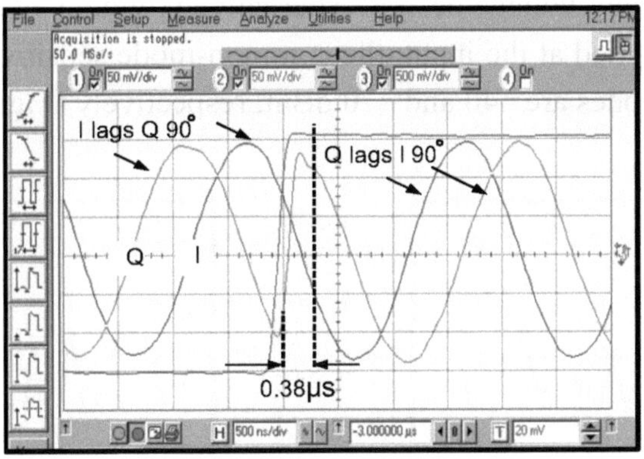

Figure 6-22. Transient measurement of IF channel selection

The channel mismatch throughout the signal band for both 802.11a and g modes are shown in Figure 6-23(a) and (b). The averaged gain/phase mismatches are 0.17 dB/0.39° and 0.16 dB/0.7° for 802.11a and g, respectively. This corresponds to an averaged IRR over the signal band of approximate 40 dB (Figure 6-24). The degradation of IRR at high frequency is dominated by phase mismatch rather than gain. The I/Q channel measured an isolation of >60 dB, and the stopband rejection ratio at 20/40 MHz offset frequency is 32/90 dB (Figure 6-25). Figure 6-26(a) shows the constancy of the gain responses for 802.11b and the gain responses for 802.11a and g at maximum gain are depicted in Figure 6-26(b). Even though there is no automatic BW-tuning circuitry included in the fabrication chip, the manually tunable BW covers 6.54–8.95 MHz. The standard deviation of the upper/lower cutoffs over 52 dB gain range is 8.6/12.4%. The spot conversion gain (~50 dB) and noise figure (~30 dB) in 1 Hz–99 kHz measured for 802.11a modes are shown in Figure 6-27. The in-band two-tone test (tones at 1.9 and 2.1 MHz) is showed in a wideband view (Figure 6-28(a)), where the noise floor in the midband and stopband are showed as well. Another midband two-tone test (tones at 3.9 and 4.1 MHz) is depicted in Figure 6-28(b), where an IM_3 of over 63 dB is measured. For the power-supply sensitivity, a –20-dBm test source applied at the ground node results in an in-band PSS^- of around –70 dBm (Figure 6-29(a)), verifying the effectiveness of the stage-dc-offset cancellation in low-voltage design. Alternatively, with a –20-dBm common-mode signal applied at the input, the common-mode sensitivity show at high and low gain modes are –40 and –70 dBm, respectively (Figure 6-29(b)).

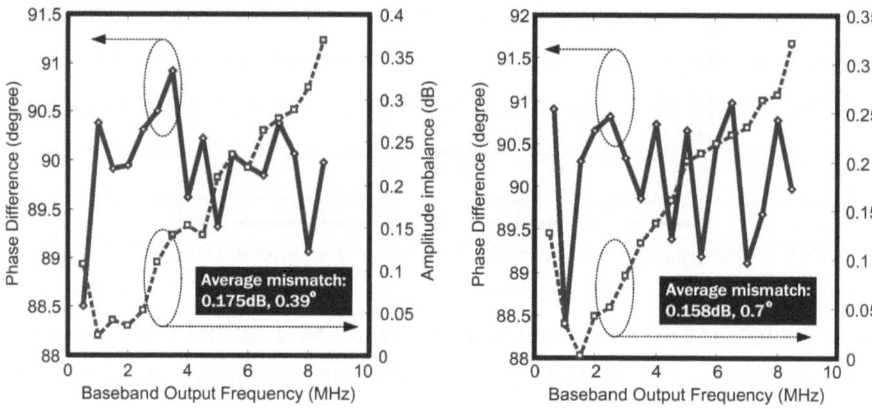

(a) (b)

Figure 6-23. I/Q mismatch over the signal band: (a) 802.11a and (b) 802.11g modes

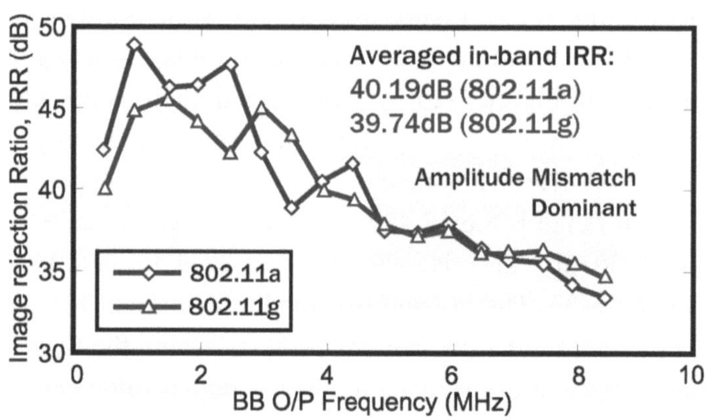

Figure 6-24. Corresponding IRR of the I/Q mismatch plot in Figure 6-23

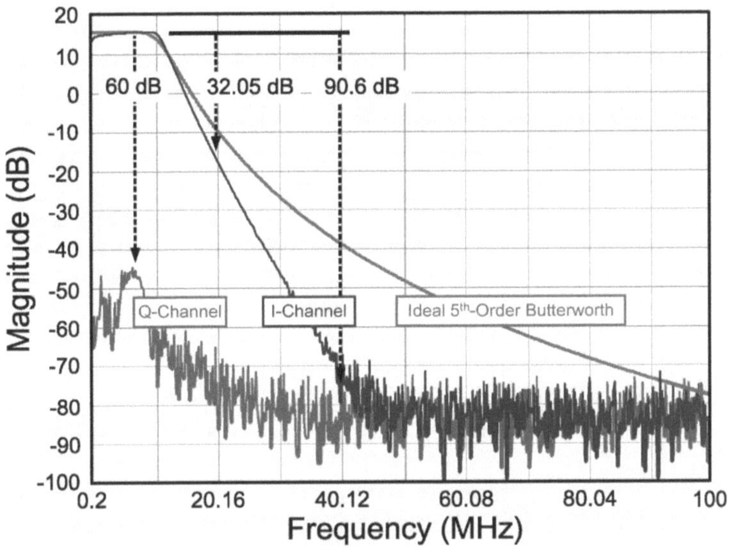

Figure 6-25. I/Q isolation and selectivity compared with a fifth-order Butterworth LPF

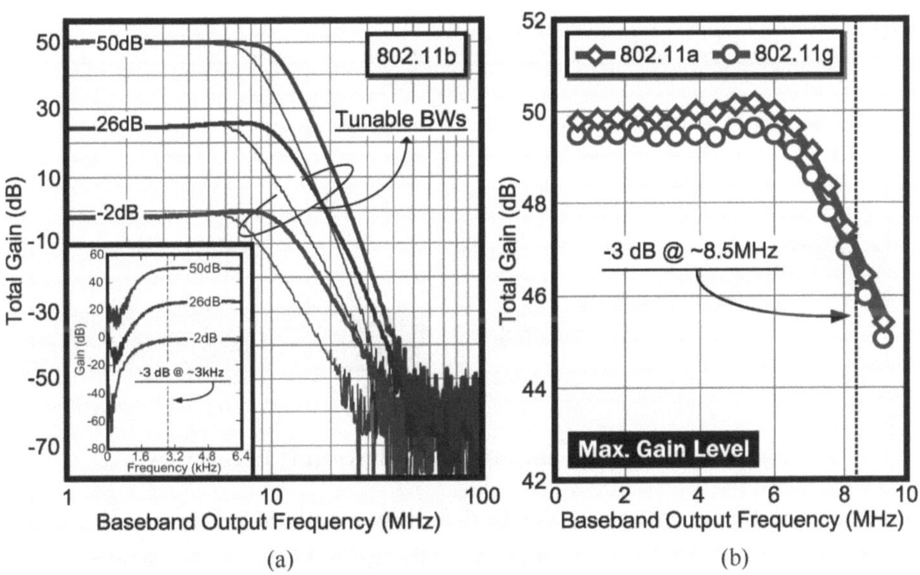

Figure 6-26. Gain responses: (a) 802.11b and (b) 802.11a and g modes [note the constant selectivity against gain in (a)]

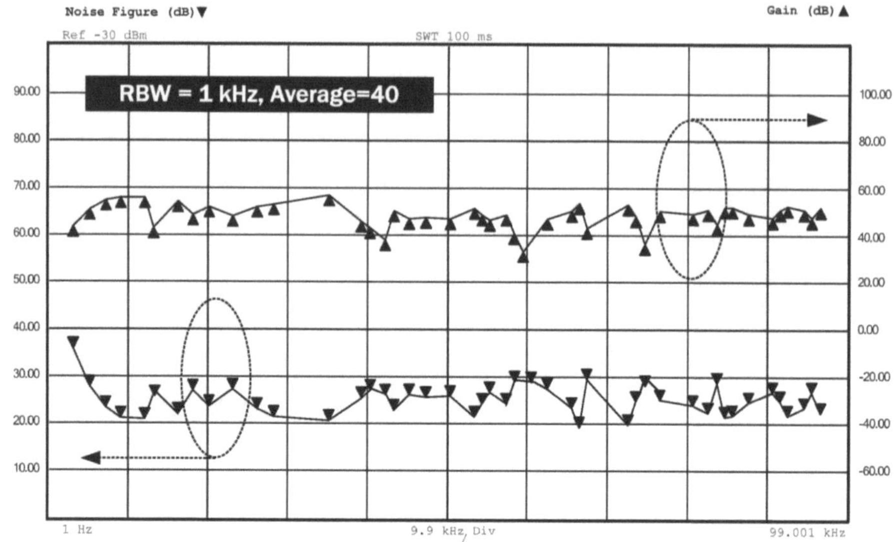

Figure 6-27. Noise figure and baseband gain in 802.11a mode

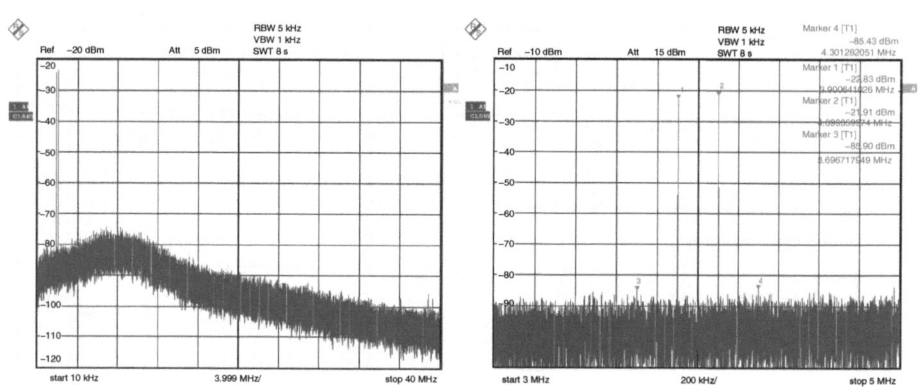

Figure 6-28. Two-tone test: (a) Wideband view (b) Close-up view

System performances were verified with constellation diagram and error vector magnitude (EVM) tests. Figure 6-30 and 6-31 show the results under a 54-Mps, –31.8-dBm, 64QAM-OFDM signal for a/g mode and an 11-Mps, –32.7-dBm, DSSS-CCK signal for b mode, respectively. Although the results (especially around dc) were partially degraded by the presence of the I/O test buffers and transformers, the former still shows an EVM of –27.03 dB (4.45%), meeting the standard allowed –25 dB (5.6%). For the latter, it demonstrates an EVM of –17.04 dB (14.07%), which again well meets the standard allowed –9 dB (35.5%).

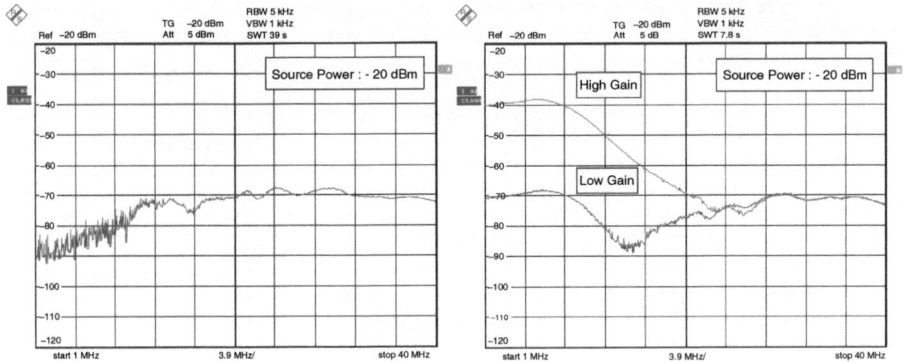

Figure 6-29. (a) Measured negative power-supply sensitivity (b) Measured common-mode signal sensitivity

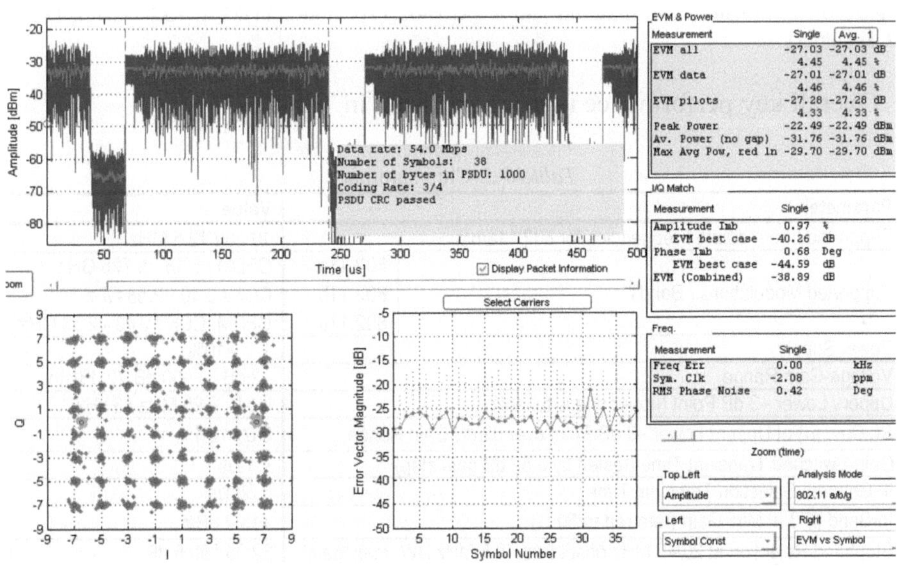

Figure 6-30. Constellation diagram and EVM (802.11a/g mode: 54-Mps, –31.8-dBm, 64QAM-OFDM signal)

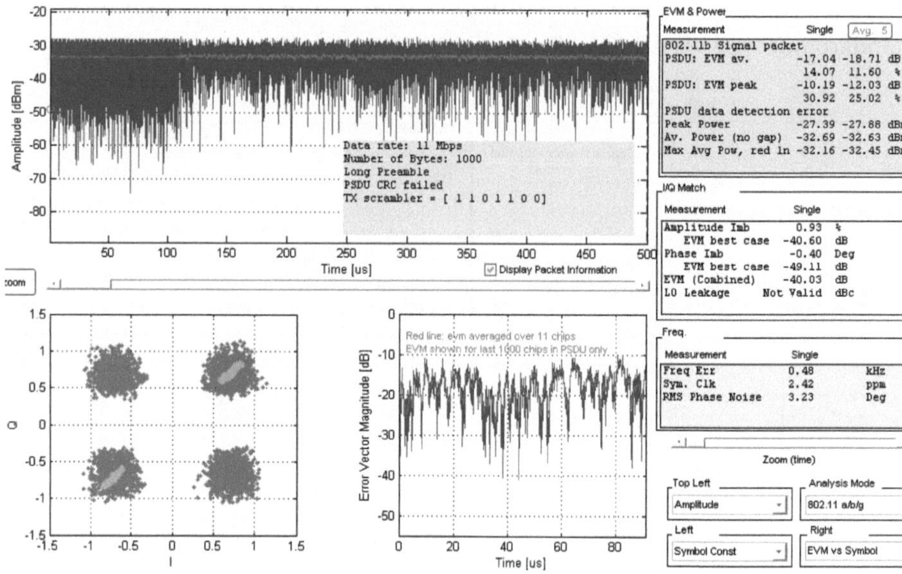

Figure 6-31. Constellation diagram and EVM (802.11b mode: 11-Mps, –32.7-dBm, DSSS-CCK signal)

Overall key performance metrics are listed in Table 6-5.

Table 6-5. Chip summary

Parameter		Value
Supported Intermediate Frequencies for 802.11a/b/g		10 / 0 / 12.5 MHz
Supported Modulations / Bands	802.11a	OFDM / 5.15 - 5.725 GHz
	802.11b	CCK / 2.40 - 2.58 GHz
	802.11g	OFDM, CCK / 2.40 - 2.58 GHz
Power Supply		1 V ±10%
Voltage-Gain Range		–2 dB ... +50 dB (2 dB/step)
Upper / Lower –3 dB Point (upper one is tunable)		6.54 - 8.95 MHz / 3 kHz
St. Dev. (σ) of Upper / Lower –3 dB Point over 52 dB Gain Range		8.6 / 12.4%
Gain-Switched Transient Time (tested by a 52 dB gain step)		< 1 μs
IF Channel-Selection Transient Time		0.38 μs
In-Band IIP3 at Min. Gain (referred to 50 Ω)		+15.2 dBm
Stopband Rejection at 20/40 MHz offset freq. (8.5 MHz BW, max. gain)		32.05 / 90.6 dB
EVM – OFDM/CCK mode		–27.03 / –17.04 dB
I/Q Isolation		> 60 dB
Averaged In-Band I/Q Impairment in 802.11a/g Mode	Amplitude	0.175 / 0.158 dB
	Phase	0.39 / 0.7°
Input-Referred white Noise Spectral Density / Noise Figure (white)		22.5 nV/√Hz / <30 dB
Power Per Channel at 0.9 / 1 / 1.1 V (excluded the test buffers)		13 / 14 † / 16.5 mW
Technology ($V_{T,n}$: 0.52 V, $V_{T,p}$: –0.65 V)		0.35 μm 4M2P CMOS
Active Core Area (I/Q PGAs + I/Q LPFs + Preselect filter & DQDC + CLK + Other Blocks)		3.06 mm^2 (1.44+1.12+0.25+0.05+0.2)

8. SUMMARY

In this chapter, the design consideration, test strategy and experimental results of a *first-silicon-success* low-voltage receiver analog-baseband IC for IEEE 802.11a/b/g WLAN have been presented. The IC fabricated in 0.35-μm CMOS operates successfully at a 1-V supply. The multistandard compliance is enabled by a flexible-IF reception, whereas the achieved 28-mW power consumption and 3-mm^2 active area for dual channels are the contributions of the proposed circuit techniques. In the characterization, the testing tools and bypassing methods that are important for high-quality and block-level measurements have been summarized.

REFERENCES

[6.1] "The International Technology Roadmap for Semiconductors – RF and Analog/ Mixed-Signal Technologies for Wireless Communications," 2004, http://www.itrs.net/ Common/2004 Update/2004_04_Wireless.pdf

[6.2] J. Arias, et al., "A 32-mW 320-MHz Continuous-Time Complex Delta-Sigma ADC for Multi-Mode Wireless-WLAN Receivers," *IEEE Journal of Solid-State Circuits (JSSC)*, vol. 41, no. 2, pp. 339–351, Feb. 2006.

[6.3] E. J. Foster, "Active Low-Pass Filter Design," *IEEE Transactions on Audio*, vol. 13, no. 5, pp. 104–111, Oct. 1965.

[6.4] A. Behzad, "Wireless LAN Radio Design," *IEEE International Solid-State Circuits Conference (ISSCC)*, Tutorial T2, Feb. 2004.

[6.5] T. Hornak, K. Knudsen, A. Grzegorek, K. Nishimura and W. McFarland, "An Image-Rejecting Mixer and Vector Filter with 55-dB Image Rejection over Process, Temperature, and Transistor Mismatch," *IEEE Journal of Solid-State Circuits (JSSC)*, vol. 36, no. 1, pp. 23–33, Jan. 2001.

[6.6] R. Baker, H. W. Li and D. E. Boyce, *CMOS Circuit Design, Layout and Simulation*, IEEE Press, pp. 528–529, 1998.

[6.7] P.-I. Mak, S.-P. U and R. P. Martins, "A 1V 14mW-per-Channel Flexible-IF CMOS Analog-Baseband IC for 802.11a/b/g Receivers," *IEEE Symposium on VLSI Circuits (VLSI), Digest of Technical Papers*, pp. 288–289, June 2006.

[6.8] P.-I. Mak, S.-P. U and R. P. Martins, "An Experimental 1-V Flexible-IF CMOS Analog-Baseband Chain for an IEEE 802.11a/b/g Wireless LAN Receiver," submitted for journal publication.

[6.9] Yamaichi Electronics, IC 149 Series Socket datasheet: Ceramic Quad Flat Package (CQFP).

[6.10] P.-I. Mak, S.-P. U and R. P. Martins, "Design and Test Strategy underlying a Low-Voltage Analog-Baseband IC for 802.11a/b/g-WLAN SiP Receivers," in *Proc. IEEE International Symposium on Circuits and Systems (ISCAS)*, pp. 2473–2476, May 2006.

[6.11] M. Montrose, EMC and the Printed Circuit Board: Design, Theory, and Layout Made Simple, IEEE Press, 1999.

[6.12] National Semiconductor, LM134/LM234/LM334:3-Terminal Adjustable Current Source Datasheet, Mar. 2000.

Chapter 7

CONCLUSIONS

1. CONCLUDING REMARKS

This book has proposed and examined various wireless transceiver architectures and circuit structures to achieve multistandard and low-voltage compliance. The key points of each chapter are summarized as follows:

- In Chapter 2, the fundamental receiver and transmitter architectures have been reviewed. The remarkable techniques reinforced in the state-of-the-art works have been highlighted. A statistical summary of the recently reported receiver and transmitter architectures for modern wireless communication systems has been surveyed, giving an updated overview of the wireless-IC evolution in the last decade.

- A coarse-RF fine-IF (two-step) channel-selection technique has been introduced in Chapter 3. The resultant advantages are, first, a relaxation of the RF frequency synthesizer and local oscillator specifications through channel-selection partitioning; and second, a synthesis of zero-IF/low-IF downconversion in the receive path and direct-up/two-step-up upconversion in the transmit path. Both architectures have been illustrated through the design of wideband-/narrowband-mixed multi-standard systems. Specific functional blocks implementing the concept have been designed and validated.

- The system design of a SiP receiver analog-baseband chain for IEEE 802.11a/b/g WLAN has been described in Chapter 4. A 3D-stack floorplan simultaneously takes the testability and routability into account. In the frequency plan, the two-step channel selection synthesizes the low-IF and zero-IF in signal reception and relaxes the channel selection at 5-GHz for 802.11a mode. A cost-efficient baseband signal conditioning approach, first, lowers the order requirement of the CSF by adopting a constant-BW PGA to increase the stopband rejection, and second, improves the overall linearity by locally adopting a dc-offset canceler at each CSF and PGA stage.

- Three novel circuit blocks have been presented in Chapter 5. The first one is a flexible-IF double-quadrature-downconversion filter, which is composed by a series-switching mixer-quad, an active-RC lowpass filter and a mixed-mode I/Q clock generator. They jointly realize a clock-rate-defined-IF I/Q demodulation and channel-selection lowpass filtering. The second one is a programmable-gain amplifier using a switched-current-resistor technique to achieve a transient-free constant-BW gain control. The third one is an inside-OpAmp dc-offset canceler, which provides switchability for the highpass poles and enhances the area efficiency in integration.

- The design of an experimental 1-V receiver analog-baseband IC for IEEE 802.11a/b/g WLAN has been presented in Chapter 6. The wideband requirement but low-voltage headroom in the available 0.35-μm CMOS technology (nMOS $V_{T,n}$ of 0.52 V and a pMOS $V_{T,p}$ of –0.65 V) have been overcome by applying a proper synthesis of low-IF and zero-IF architectures, and a series of low-voltage-enabled circuits that have been presented in Chapter 5. The test strategy enables both block-level and system-level measurements.

2. BENCHMARKS

2.1 Functional-block level

Comparing with the previously reported works [7.2][7.3][7.4][7.5][7.6][7.7][7.8][7.9][7.10][7.11][7.12][7.13][7.14][7.15], this work, the PGA [7.1], utilizes the lowest supply voltage reported up to date while giving a medium tunable gain range of 52 dB (Figure 7-1).

On the other hand, comparing with the various low-voltage (≤1.5 V) CT analog functional blocks including lowpass filters [7.16][7.17][7.18][7.19] and sigma-delta (ΣΔ) modulators [7.20][7.21][7.22][7.23][7.24][7.25][7.26], the proposed filter, DQDC, and PGA stay at low-voltage high-frequency/-bandwidth positions (Figure 7-2) without using a leading-edge or specialized process.

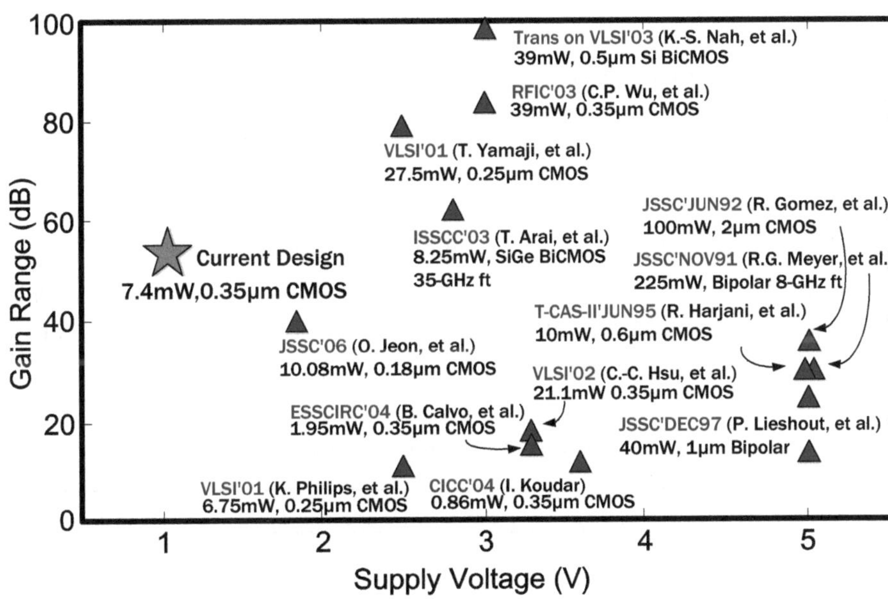

Figure 7-1. Brief comparison of several state-of-the-art baseband CT-PGAs & CT-GCAs

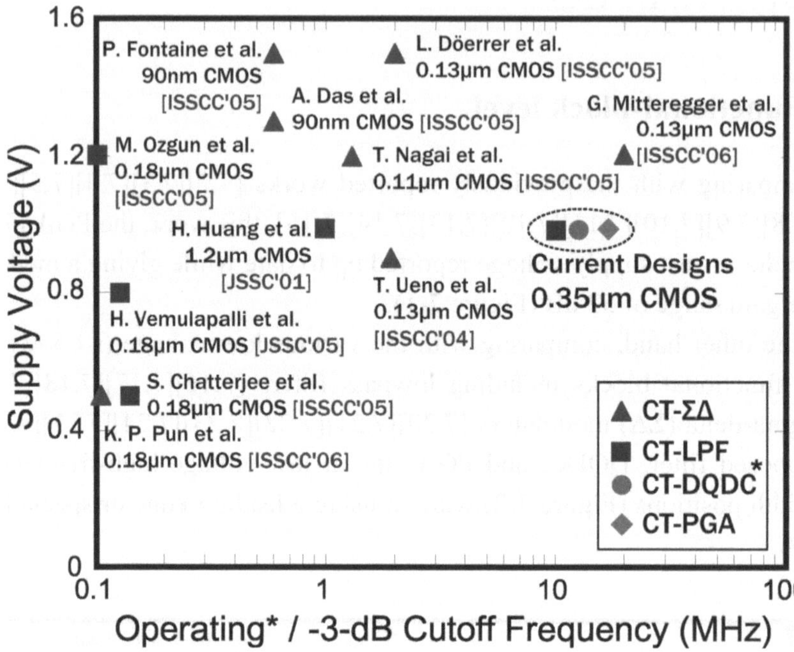

Figure 7-2. Brief comparison of several state-of-the-art low-voltage CT baseband functional blocks

2.2 Subsystem level

Comparing with the previously reported baseband chains [7.27][7.28][7.29][7.30][7.31], the current design [7.32] employs the lowest supply voltage reported up to date without sacrificing the power, noise and linearity, or relying on leading-edge/specialized technologies (Table 7-1).

Table 7-1. Brief comparison of state-of-the-art analog-baseband chains

	J. Jussila *et al.* ESSCIRC'99 [7.27]	W. Schelmbauer *et al.* RFIC'02 [7.28]	H. O. Elwan *et al.* TCAS-II'02 [7.29]	M. Lee *et al.* ESSCIRC'04 [7.30]	M. Elmala *et al.* VLSI'05 [7.31]	The current Design VLSI'06 [7.32]
Applications	WCDMA	WCDMA	Bluetooth	ZigBee	802.11a/b/g WLANs	802.11a/b/g WLAN
Supply Voltage	2.7 V...3 V	2.7 V	3 V	1.8 V	1.4 V	1 V
Power per Channel	58.5 mW	19.4 mW	3.6 mW	2.43 mW	13.5 mW	14 mW
Input-referred Noise/Noise Density /Noise Figure (NF)	11 μV_{rms} (integrated from 100 Hz... 20 MHz)	8 nV/√Hz	43.2 nV/√Hz	31 dB NF	19 nV/√Hz	22.5 nV/√Hz
IIP3 (minimum gain)	+14 dBm	+20.6 dBVrms	+12.2 dBm	+30 dBm	+2 dBm	+15.2 dBm
Stopband attenuation	N/A	43 dB @ (3.1-6.9 MHz)	62 dB @ 15 MHz	72 dB @ 5 MHz	N/A	32/90 dB @ 20/40 MHz
Gain Range	-9... +69 dB	-15.5... +48.5 dB	+12... +30 dB	+12... +55 dB	+13.5... +67.5 dB	-2... +50 dB
Technology	0.35 μm BiCMOS	75 GHz SiGe BiCMOS	1.2 μm CMOS	0.18 μm CMOS	90 nm CMOS	0.35 μm CMOS

3. RECOMMENDATIONS FOR FUTURE WORK

There are numerous issues awaiting further exploration in future research work:

- At the architectural level, the disclosed low-IF/zero-IF-reconfigurable solution for IEEE 802.11a/b/g can be further extended to the upcoming 802.11n in two different aspects. The first one is due to the required dynamic bandwidth reception. On top of the traditional 20-MHz mode, a new 40-MHz mode combined two 20-MHz channels on 108 OFDM, is required to support it. Obviously, this new feature implies the channel center and BW change together between zero-IF (at 40-MHz mode) and low-IF (at 20-MHz mode). Using dual flexible-IF analog basebands with a fixed-BW of 20 MHz to process the upper and lower sidebands separately appears as a fast-to-develop solution.

  On the other hand, 802.11n also supports spatial multiplexing multiple-input multiple-output (MIMO) communications. With a 2x2 structure, for instance, the dual analog basebands can be used to support two independent MIMO paths. Combining these two concerns, a timely future work will consist in developing a flexible-IF analog-baseband platform for IEEE 802.11a/b/g/n-compliant receivers.

- Based on the architectures and low-voltage techniques investigated in this book, a low-voltage transmitter analog baseband can be developed correspondingly to complement the receiver path.

- Among the proposed circuit techniques, the inside-OpAmp dc-offset cancellation is generally applicable to many kinds of differential circuits for area savings. For the switched-current-resistor PGA, it is capable to work technology-independently under a supply voltage of just 0.6 V, sufficiently fulfilling the voltage scaling trend in the upcoming technologies. On the other hand, among the transceiver analog baseband, there are still certain miscellaneous analog-baseband circuitry such as the receiver signal strength indicator (RSSI), the automatic time-constant tuning circuit, the I/Q-mismatch calibration circuit and the low-dropout bandgap reference, that are needed. Realizing them under a low-voltage supply still remains as a challenging task.

- For multistandard applications, the radio front-end circuits such as wideband (DC-to-6 GHz) low-noise amplifier, quadrature mixer, frequency synthesizer, local oscillator and power amplifier constitute great challenges of designing under low-voltage constraints. Design of low-voltage multimode radio front-end would be a very interesting and challenging topic.

- At the radio's back end, multistandard transceivers require the A/D and D/A interfaces to cover a wide range of speed and resolution requirements. For power and area savings, design of low-voltage power-efficient A/D and D/A interfaces with scalable speed and resolution would be very imperative. The potential A/D structures are pipelined, successive approximation and CT-ΣΔ, whereas it is the current-steering D/A structure that is befitting multistandard transmitter for its inherent clock-rate definability in conversion speed.

- Adaptive performance-on-demand control is also a new direction to lower the average power consumption of WLAN systems. The analog front-end's noise and linearity performances are managed dynamically by tracking the incoming signal power and its nearby interferer's strength.

REFERENCES

[7.1] P.-I. Mak, S.-P. U and R. P. Martins, "A 1-V Transient-Free and DC-Offset-Canceled PGA with a 17.1-MHz Constant Bandwidth over 52-dB Control Range in 0.35-μm CMOS," in *Proc. IEEE Custom Integrated Circuits Conference (CICC)*, pp. 649–652, Sept. 2005.

[7.2] R. G. Meyer and W. D. Mack, "A DC to 1-GHz Differential Monolithic Variable-Gain Amplifier," *IEEE Journal of Solid-State Circuits (JSSC)*, vol. 26, no. 11, pp. 1673–1680, Nov. 1991.

[7.3] R. Gomez and A. A. Abidi, "A 50-MHz CMOS Variable-Gain Amplifier for Magnetic Data Storage Systemx," *IEEE Journal of Solid-State Circuits (JSSC)*, vol. 27, no. 6, pp. 935–939, June 1992.

[7.4] R. Harjani, "A Low-Power CMOS VGA for 50Mb/s Disk Drive Read Channels," *IEEE Transactions on Circuits and Systems-II (TCAS-II)*, vol. 42, pp. 370–376, June 1995.

[7.5] J. J. F. Rijns, "CMOS Low-Distortion High-Frequency Variable-Gain Amplifier," *IEEE Journal of Solid-State Circuits (JSSC)*, vol. 31, no. 7, pp. 1029–1034, July 1996.

[7.6] P. J. G. van Lieshout and R. J. van de Plassche, "A Power-Efficient, Low-Distortion Variable Gain Amplifier Consisting of Coupled Differential Pairs," *IEEE Journal of Solid-State Circuits (JSSC)*, vol. 32, no. 12, pp. 2105–2110, Dec. 1997.

[7.7] T. Yamaji, N. Kanou and T. Itakura, "A Temperature-Stable CMOS Variable-Gain Amplifier With 80-dB Linearly Controlled Gain Range," *IEEE Symposium on VLSI Circuits (VLSI), Digest of Technical Papers*, pp. 77–80, 2001.

[7.8] K.-S Nah and B.-H Park, "A 50-MHz 98-dB Dynamic-Range dB-Linear Programmable-Gain Amplifier with 2-dB Gain Steps for 3-V Power Supply," *IEEE Symposium on VLSI Circuits (VLSI), Digest of Technical Papers*, pp. 73–76, 2001.

[7.9] K. Philips and E. C. Dijkmans, "A Variable-Gain IF Amplifier with –67dBc IM3-distortion at 1.4 V_{pp} Output in 0.25 μm CMOS," *IEEE Symposium on VLSI Circuits (VLSI), Digest of Technical Papers*, pp. 81–82, June 2001.

[7.10] C.-C. Hsu and J.-T. Wu, "A 125 MHz –86 dB IM3 Programmable-Gain Amplifier," *IEEE Symposium on VLSI Circuits (VLSI), Digest of Technical Papers*, pp. 32–35, June 2002.

[7.11] T. Arai and T. Itakura, "A Baseband Gain-Controlled Amplifier with a Linear-in-dB Gain Range from 14dB to 76dB and a Fixed Corner Frequency DC Offset Canceler," *IEEE International Solid-State Circuits Conference (ISSCC), Digest of Technical Papers*, pp. 136–137, Feb. 2003.

[7.12] C.-P. Wu and H.-W. Tsao, "A 110 MHz 84 dB CMOS programmable gain amplifier with RSSI," *IEEE Radio Frequency Integrated Circuits Symposium (RFIC), Digest of Technical Papers*, pp. 639–642, June 2003.

[7.13] I. Koudar, "Variable Gain Differential Current Feedback Amplifier," in *Proc. IEEE Custom Integrated Circuits Conference (CICC)*, pp. 659–662, Oct. 2004.

[7.14] B. Calvo, S. Celma and M. T. Sanz, "A High-Linear 160-MHz CMOS PGA, in *Proc. European Solid-State Circuits Conference (ESSCIRC)*, pp. 115–118, Oct. 2004.

[7.15] O. Jeon, R. M. Fox and B. A. Myers, "Analog AGC Circuitry for a CMOS WLAN Receiver," *IEEE Journal of Solid-State Circuits (JSSC)*, vol. 41, no. 10, pp. 2291–2300, Oct. 2006.

[7.16] H. Huang and E. K. F. Lee, "Design of Low-Voltage CMOS Continuous-Time Filter with On-Chip Automatic Tuning," *IEEE Journal of Solid-State Circuits (JSSC)*, vol. 36, no. 8, pp. 1168–1177, Aug. 2001.

[7.17] M. Ozgun, Y. Tsividis and G. Burra, "Dynamically Power-Optimized Channel-Select Filter for Zero-IF GSM," *IEEE International Solid-State Circuits Conference (ISSCC), Digest of Technical Papers*, pp. 504–505, Feb. 2005.

[7.18] S. Chatterjee, Y. Tsividis and P. Kinget, "A 0.5V Filter with PLL-Based Tuning in 0.18μm CMOS," *IEEE International Solid-State Circuits Conference (ISSCC), Digest of Technical Papers*, pp. 506–507, Feb. 2005.

[7.19] G. Vemulapalli, P. K. Hanumolu, Y.-J. Kook and U. K. Moon, " A 0.8-V Accurately Tuned Linear Continuous-Time Filter," *IEEE Journal of Solid-State Circuits (JSSC)*, vol. 40, no. 9, pp. 1972–1977, Sept. 2005.

[7.20] T. Ueno and T. Itakura, "A 0.9V 1.5mW Continuous-Time ΔΣ Modulator for WCDMA," *IEEE International Solid-State Circuits Conference (ISSCC), Digest of Technical Papers*, pp. 78–79, Feb. 2004.

[7.21] L. Dorrer, F. Kuttner, P. Greco and S. Derksen, "A 3mW 74dB SNR 2MHz CT ΔΣ ADC with a Tracking-ADC-Quantizer in 0.13μm CMOS," *IEEE International Solid-State Circuits Conference (ISSCC), Digest of Technical Papers*, pp. 492–493, Feb. 2005.

[7.22] T. Nagai, H. Satou, H. Yamazaki and Y. Watanabe, "A 1.2V 3.5mW ΔΣ Modulator with a Passive Current Summing Network and a Variable Gain Function," *IEEE International Solid-State Circuits Conference (ISSCC), Digest of Technical Papers*, pp. 494–495, Feb. 2005.

[7.23] P. Fontaine, A. N. Mohiedin and A. Bellaouar, "A Low-Noise Low-Voltage CT ΔΣ Modulator with Digital Compensation of Excess Loop Delay," *IEEE International Solid-State Circuits Conference (ISSCC), Digest of Technical Papers*, pp. 498–499, Feb. 2005.

[7.24] A. Das, R. Hezar, R. Byrd, G. Gornez and B. Haroun, "A 4th-Order 86dB CT ΔΣ ADC with Two Amplifiers in 90nm CMOS," *IEEE International Solid-State Circuits Conference (ISSCC), Digest of Technical Papers*, pp. 496–498, Feb. 2005.

[7.25] G. Mitteregger, C. Ebner, S. Mechnig, T. Blon, C. Holuigne, E. Romani, A. Melodia and V. Mellimi, "A 14b 20mW CMOS CT ΔΣ ADC with 20MHz Signal Bandwidth and 12b ENOB," *IEEE International Solid-State Circuits Conference (ISSCC), Digest of Technical Papers*, pp. 62–63, Feb. 2006.

[7.26] K.-P. Pun, S. Chatterjee and P. Kinget, "A 0.5V 74dB SNDR 25 kHz CT ΔΣ Modulator with Return-to-Open DAC," *IEEE International Solid-State Circuits Conference (ISSCC), Digest of Technical Papers*, pp. 72–73, Feb. 2006.

[7.27] J. Jussila A. Pärssinen and K. Halonen, "An Analog Baseband Circuitry for a WCDMA Direct Conversion Receiver," in *Proc. European Solid-State Circuits Conference (ESSCIRC)*, pp. 166–169, Sept. 1999.

[7.28] W. Schelmbauer, H. Pretl, L. Maurer, B. Adler, R. Weigel, R. Hagelauer and J. Fenk, "An Analog Baseband Chain for UMTS Zero-IF Receiver in a 75GHz SiGe BiCMOS Technology," *IEEE Radio Frequency Integrated Circuits Symposium (RFIC), Digest of Technical Papers,* pp. 267–270, June 2002.

[7.29] H. Elwan, M. Younus, H. Al-Zaher and M. Ismail, "A Buffer-Based Baseband Analog Front End for CMOS Bluetooth Receivers," *IEEE Transactions on Circuits and Systems-II (TCAS-II)*, vol. 49, no. 8, pp. 545–554, Aug. 2002.

[7.30] M. Lee, I. Kwon and K. Lee, "An Integrated Low Power CMOS Baseband Analog Design for Direct Conversion Receiver," in *Proc. European Solid-State Circuits Conference (ESSCIRC)*, pp. 79–82, Sept. 2004.

[7.31] M. Elmala, B. Carlton, R. Bishop and K. Soumyanath, "A Highly Linear Filter and VGA Chain with Novel DC-Offset Correction in 90nm Digital CMOS Process," *IEEE Symposium on VLSI Circuits (VLSI), Digest of Technical Papers*, pp. 302–303, June 2005.

[7.32] P.-I. Mak, S.-P. U and R. P. Martins, "A 1V 14mW-per-Channel Flexible-IF CMOS Analog-Baseband IC for 802.11a/b/g Receivers," *IEEE Symposium on VLSI Circuits (VLSI), Digest of Technical Papers*, pp. 288–289, June 2006.

MIX
Papier aus verantwortungsvollen Quellen
Paper from responsible sources
FSC® C105338

If you have any concerns about our products,
you can contact us on
ProductSafety@springernature.com

In case Publisher is established outside the EU,
the EU authorized representative is:
Springer Nature Customer Service Center GmbH
Europaplatz 3, 69115 Heidelberg, Germany

Printed by Libri Plureos GmbH
in Hamburg, Germany